Forest and Plant Bioproducts Division 2018

Core Programming Area at the 2018 AIChE Annual Meeting

Pittsburgh, Pennsylvania, USA
28 October - 2 November 2018

ISBN: 978-1-5108-7618-7

Printed from e-media with permission by:

Curran Associates, Inc.
57 Morehouse Lane
Red Hook, NY 12571

Some format issues inherent in the e-media version may also appear in this print version.

Printed by Curran Associates, Inc. (2019)

AIChE
120 Wall Street, FL 23
New York, NY 10005-4020

Phone: (800) 242-4363
Fax: (203) 775-5177

www.aiche.org

Additional copies of this publication are available from:

Curran Associates, Inc.
57 Morehouse Lane
Red Hook, NY 12571 USA
Phone: 845-758-0400
Fax: 845-758-2633
Email: curran@proceedings.com
Web: www.proceedings.com

TABLE OF CONTENTS

Some abstracts in this book refer to images that were not made available by the publisher

532921 Ultrapure Lignin Via the ALPHA Process for Materials Applications: From Carbon Fibers to Coatings

Abstract:

Among renewable feedstocks, lignin can play an important role as a precursor for carbon fibers, owing to its significant carbon content. Unfortunately, the inexpensive, commercial-grade lignins available today (primarily Kraft lignins) are too polydisperse with a high metals/ash content, eliminating them from contention for high-value

Thies and co-workers have developed a process for simultaneously solvating, fractionating, and purifying the lignin polymer recovered from biomass by-product streams of pulp-and-paper mills and lignocellulosic biorefineries. Aqueous Lignin Purification with Hot Acids (ALPHA) involves combining solid lignin with hot acetic acid–water solvent mixtures to produce two liquid phases: a highly solvated, (lignin) polymer-rich phase and a solvent-rich phase. ALPHA can be operated in two ways: (1) by using increasingly aggressive solvent mixtures to isolate ultrapure, metals-free (i.e., <100 ppm) lignin fractions of medium/high molecular weight (MW), or (2) by "reversing" the process and using solvent mixtures of decreasing strength to isolate metals-free lignin fractions of low molecular weight (MW).

ALPHA was used to generate ultrapure, high MW fractions of lignin that were then converted into carbon fibers; the resulting tensile strengths and moduli were almost doubled than any reported to date. Furthermore, "reverse" ALPHA was used to generated ultrapure, low MW fractions of lignin for application in coatings. Also discussed in this work will be how one can tune the parameters of ALPHA (e.g., temperature, acetic acid/water ratio, solvent-to-lignin ratio) to generate lignin fractions with the properties (including phenolic content, molecular weight, and purity) required for specific applications.

529076 Mixture-Process Variable Experimental Design to Optimize Sugar Mixture (glucose, xylose and arabinose) Conversion to Polyhydroxybutyrate By *Burkholderia Saccharia*

Abstract:

Polyhydroxybutyrate (PHB) is a biodegradable plastic that is comparable with some petroleum based plastics in terms of mechanical properties. Conversion of corn stover or corn fiber to PHB makes corn processing more profitable and sustainable. Glucose, xylose and arabinose are the main monomer sugars derived from either corn stover or corn fiber. *Burkholderia sacchari* DSM 17165 is a bacteria strain that can convert all three sugars into PHB. However, the effects of sugar ratio in mixture, process variable (sugar concentration, carbon to nitrogen ratio) and their interactions on PHB concentration and productivity is seldom studied. In this study, mixture-process variable experimental design is employed. Single factor experimental results showed that the optimal initial sugar concentrations and carbon to nitrogen molar ratios ranged from 10 to 40 g/L and 40 to 80 for different sugar mixture ratios. A ten run design for sugar ratio combined with a full factorial design for optimizing process variables will be performed to optimize PHB production.

540653 Integrating Sustainable Biocarbon in Lightweight and Durable Biocomposite Solutions for Automotive Applications

Abstract:

Due to modern efforts to curb pollution through policy and regulations, the reduction of greenhouse gas (GHG) emissions has become a critical aspect for new innovation in the automotive sector. The most simple solution for this issue is to reduce the weight of new vehicles, thus reducing fuel consumption. However, it is of paramount importance to reduce weight without compromising safety or performance. Natural fibre reinforced composites have been investigated and adopted by several manufacturers in recent years. Unfortunately, these materials have several flaws, such as their poor thermal stability, and processing odours, which prevent their use in engineering thermoplastics. Engineering thermoplastics such as polyamides are necessary for automotive components which will be subjected to high temperatures, pressures and stresses.

In this research, hybrid carbonaceous biocomposites of nylon based blends have been reinforced with combinations of biocarbon and other secondary fillers used to tailor various aspects. Biocarbon is a renewable, sustainable filler with excellent thermal stability suited for polyamide resins. Biocarbon is economical and also presents a green solution, as it is a carbon-neutral material. Through hybridization with other additives and heat ageing additives properties such as tensile modulus, thermal stability, heat deflection temperature, and coefficient of linear thermal expansion (CLTE) and durability can be precisely tailored. These hybrid biocomposites can be used as drop-in resins for injection moulding applications, and can have up to a 15% reduction in weight compared to conventional glass fibre/mineral filled nylon composites.

Acknowledgements

This research is financially supported by the Ontario Ministry of Agriculture, Food and Rural Affairs (OMAFRA), Canada/University of Guelph-Bioeconomy for Industrial Uses Research Program Theme (Project # 030176); the Natural Sciences and Engineering Research Council (NSERC), Canadaâ˘'Discovery Grants (Project # 401111); Agriculture and Agri-Food Canada and Competitive Green Technologies (CGTech) through AgriInnovation Program (Project # 053407); and the Ontario Research Fund, Research Excellence Program; Round-7 (ORF-RE07) from the Ontario Ministry of Research and Innovation (MRI), currently known as the Ontario Ministry of Research, Innovation and Science (MRIS) (Project # 052665 and #052644).

524510 Nano-Engineered Cement Combining Biomass Ash with Nanoparticles

Abstract:

The production of Ordinary Portland Cement (OPC) causes greenhouse gas emissions, which are a major environmental issue. Finding ways to replace OPC with other materials is essential to maintaining our worldâ€™s ecosystems. Fly ash from burning coal can be used as a replacement, but it contains hazardous substances and is becoming less available. Byproducts from forests or agricultural production are more sustainable and renewable as substitutes for OPC. Rice husks, a waste product, are removed from the rough grain at a processing center, so that they do not need to be separately transported from the fields. When combusted using the proper temperature regimen, rice husks can produce energy and a high silica ash. In the present work, we have incorporated this ash and other nano-pozzolans (OPC substitutes) into cement and have found combinations that substantially increase compressive strength. The microscale properties of these cements were also investigated to find the underlying reasons for the enhancement of properties.

540748 A Study on the Gelation Kinetics and Chain Relaxation of Polybutylene Succinate (PBS) By Reactive Extrusion

Abstract:

Reactive extrusion has been applied to fabricate high compatibility blends with enhanced performance for decades. However, controlling the gelation of polymers in reactive extrusion is crucial to achieve stable processing and high-quality products. As a biodegradable polymer, polybutylene succinate (PBS) is likely to be attacked by the free radicals at high temperature, leading to chain branching or grafting with other functional groups to prepare the modified biomaterials. In this work, the peroxide concentration and temperature dependence of PBS critical gel kinetics and viscoelastic properties were determined by rheological studies. The gel time, relaxation exponent of the critical gel and activation energy for the gelation were also determined. As a result, it was found that the gelation is a complex reaction order was 0.67, which indicated a first-order reaction occurred with additional side reactions such as chain branching or decomposition. The relaxation exponent decreased as peroxide concentrations and temperature increased, suggesting that the cross-linking density of the critical gels increased during the real reactive extrusion. At 0.5 phr peroxide, the active energy reaches a maximal value of ~93.7 kJ/mol, which indicated there were difficulties within the relaxation of the reaction cross-link network, and should therefore be avoided during melt processing. The whole research demonstrates a theoretic predication on controlling the gelation and viscoelastic properties of bio-PBS in reactive extrusion.

Acknowledgment: This research is financially supported by the Ontario Research Fund, Research Excellence Program; Round-7 (ORF-RE07) from the Ontario Ministry of Research, Innovation and Science (MRIS) (Project # 052644 and 052665); the Ontario Ministry of Agriculture, Food and Rural Affairs (OMAFRA)/University of Guelph Gryphonâ €™s LAAIR Program (Project # 298636); Agriculture and Agri-Food Canada (AAFC) and Competitive Green Technologies through AgriInnovation Program project (Project # 052882 and 051910); and the Natural Sciences and Engineering Research Council (NSERC), Canada Discovery Grants (Project # 401111 and 400320).

540750 Chicken Feather Biocarbon Based Novel Biodegradable Composites

Abstract:

Waste Chicken Feather Biocarbon Based Sustainable Composites

Zonglin Li[1], Christoff Reimer[2], **Amar Mohanty[1,2]**, Manjusri Misra[1,2]

[1]Bioproducts Discovery and Development Center, Department of Plant Agriculture, Crop Science Building, University of Guelph, Guelph, N1G 2W1, Ontario, Canada

[2]School of Engineering, Thornbrough Building, University of Guelph, Guelph, N1G 2W1, Ontario, Canada

Abstract: This work describes the conversion of chicken feathers (ChF) into biocarbon by using slow pyrolysis. The effects of pyrolysis temperature on the yield content, ash content, functionality, elemental composition, as well as thermal and electrical conductivity of the biocarbon were studied. ChF biocarbon was also incorporated with poly(lactic acid) (PLA) and the properties of PLA biocomposites were also investigated in detail. It was found that with increasing the pyrolysis temperature, the yield content of ChF biocarbon decreased. The ash content, measured by both gravimetric method and thermal gravimetric analysis (TGA) methods, increased in biocarbon samples pyrolyzed at higher temperatures. FTIR results showed that almost all of the functional groups were removed after pyrolysis at 600 °C. The elemental composition results indicated that the carbon (C) content increased with increasing pyrolysis temperature, while oxygen (O) and sulfur (S) content decreased. XRD results indicated that the carbon structure became more disordered with increasing pyrolysis temperatures. The electrical conductivity and thermal conductivity was found to decrease with the increase in the pyrolysis temperature. In the PLA/ChF composites, ChF biocarbon have proven to effectively increase the modulus of the PLA composites. This work provides a potential method for utilization of waste ChF.

Acknowledgements:

This research is financially supported by the Ontario Research Fund, Research Excellence Program; Round-7 (ORF-RE07) from the Ontario Ministry of Research, Innovation and Science (MRIS) (Project #052644 and 052665); and the Natural Sciences and Engineering Research Council (NSERC), Canada Discovery Grants Project # 401111 and 400320. We also acknowledge Maple Leaf Foods, Canada for providing clean chicken feather samples to carry out this research.

Reducing Energy Consumption in Thermomechanical Pulp Production Using Chlorine Dioxide

Rodger Beatson[1], Jayg Dimayacyac[2], Zhaoyang Yuan[2]*

[1]*British Columbia Institute of Technology, Burnaby, BC Canada V5G 3H2, Email:* rodger_beatson@bcit.ca

[2]*Department of Chemical & Biological Engineering, University of British Columbia, Vancouver, BC Canada V6T 1Z4*

ABSTRACT

Hemlock and pine/spruce high-freeness pulps have been treated with chlorine dioxide over a range of temperatures (25-65 °C), times (10-120 min) and chlorine dioxide charges (0.5-5% w/w). These chlorine dioxide treatments were followed by alkali soaking or reaction with alkaline peroxide. The resulting changes in pulp properties were determined.

For the hemlock furnish, a small but rapid increase in tensile strength occurs under mild chlorine dioxide treatment conditions. Treatment with 1% chlorine dioxide for 10 min at 45 °C and 10% consistency increased the tensile index by 3 N·m/g. A subsequent alkali soak at room temperature and 4% consistency with a charge of 0.4% w/w of sodium hydroxide on pulp increased the tensile by another 10 N·m/g giving pulp with a final tensile index around 40 N·m/g. Pulp brightness was lowered by the chlorine dioxide treatment from 55% ISO to 45% ISO and even further to 36% ISO on soaking with alkali. The dark chlorine dioxide treated pulp could be easily brightened with highly alkaline peroxide to give a final brightness of 72% ISO and a high tensile of 42 N·m/g.

For the pine/spruce furnish, little change in tensile was observed on chlorine dioxide treatment, but again a subsequent 1% sodium hydroxide soak increased the tensile strength. For both 1% and 5% w/w chlorine dioxide treatments for 10 min the tensile strength increased by 7 N·m/g. For the 5% chlorine dioxide treatment only, a further increase in tensile strength of 7 N·m/g was obtained by increasing the alkali loading to 3%. The overall gain in tensile was 78% from 18 N·m/g to 32 N·m/g. As with the hemlock furnish, the brightness dropped on chlorine dioxide treatment, although with this furnish, no further drop in brightness was seen after the alkali soak.

Both untreated and treated pine/spruce pulps were refined in a "Waring Blendor" to determine the effects of chlorine dioxide treatment on the refining of the pulp. Pulp was treated with 1% chlorine dioxide followed by a 1% alkali soak. For untreated pulp, 30 min of refining increased tensile strength by 11.8 N·m/g. A better response to refining was observed for the treated pulp, with 30 min refining increasing the tensile strength by 15.6 N·m/g to give a final tensile of 41 N·m/g. It was estimated that the energy required to obtain a tensile of 30 N·m/g could be reduced by 253kWh/t. Refining decreased the freeness of untreated pulp by up to 304 mL and the freeness of treated pulp by up to 365 mL. In both cases, refining increased scattering coefficient.

The darkening of the pulp during chlorine dioxide treatment is attributed to the formation of quinones, a pink colour being observed. Such chromophores are readily removed by the hydroperoxyl anion during alkaline peroxide treatment. A positive correlation was found between tensile gains and acid group content of the pulp, implying that the formation of carboxylic acid in the fines and fibres creates increased bonding. The gain in tensile on alkaline soaking is attributed to ionization of the acid groups and swelling of the fibres and fines.

Yield losses were estimated by chemical oxygen demand (COD) measurements on the filtrate from the treatments. The chlorine dioxide treatments caused a small yield loss of around 1%. An additional yield drop of 4% occurred on the alkaline peroxide bleaching. Sodium hydroxide soaking of the pulp at room temperature had little effect on yield loss.

INTRODUCTION

There is an increasing demand for wood fibre products that could be met, in part, by greater use of high yield pulps such as thermomechanical pulp (TMP). However, compared to bleached chemical pulp, TMP generally has lower strength properties, lower brightness and exhibits brightness reversion; this limits further utilization of TMP in the production of high-grade products. In addition, the TMP process consumes a large amount of electrical energy. A promising strategy to overcome these drawbacks could be conducting chemical treatment on the high-consistency (HC) refined TMP prior to low-consistency (LC) refining.

In prior work, we have shown that the treatment of TMP with oxidizing agents such as, highly alkaline hydrogen peroxide, oxygen and ozone, can improve the tensile strength of TMP and protect the fibre from cutting during subsequent LC refining (1-4). We estimate that the combination of oxidation and LC refining can result in total energy savings, to a given tensile, of around 900 kWh/t. The improved tensile on oxidative treatments is a result of the formation of acid groups in the fibres and fines. These high-strength "mechanical pulps" could open up opportunities for further replacement of low yield chemical pulps in markets such as high-grade printing papers and packaging. With this in mind, it was of interest to explore the potential of using chlorine dioxide as an oxidizing/strengthening agent for mechanical pulp. This would be a particular interest to mills that produce both bleached kraft and mechanical pulps at the same site.

Chlorine dioxide is a strong oxidizing agent commonly used in kraft pulp bleaching. It was originally introduced in first stage chlorine bleaching as a means of increasing pulp strength (5). As the environmental concerns grew over the presence of dioxin and chlorinated aromatics in bleach plant effluents it was recognised that the levels of chlorinated organics could be drastically reduce by increased substitution of chlorine with chlorine dioxide. In almost all bleached kraft pulp mills, chlorine dioxide has completely replaced chlorine for strength and environmental reasons leading to Absorbable Organic Halide (AOX) levels below 0.7 kg/ton (6). Chlorine dioxide mainly oxidises the lignin at free-phenolic structures to generate carboxylic (muconic) acids. Chlorine dioxide, unlike chlorine, in not capable of performing electrophilic substitution of the hydrogen on the aromatic rings which leads to the chlorinated aromatics (7). In the late 1970s some early work showed the potential of using chlorine dioxide in a two-stage oxidation and bleaching process on chemimechanical pulps (8, 9).

In this work, we report the results of an investigation of the effect of chlorine dioxide treatments of mechanical pulps on the physical and optical properties before and after an alkali soak, subsequent lab-scale low consistency refining and brightening. Yields of the treatments were estimated by measurement of Chemical Oxidation Demand (COD)(10).

EXPERIMENTAL

Raw Material

Hemlock and spruce/pine high-freeness thermomechanical pulps with freeness values of 380 mL and 680 mL were obtained from Catalyst Paper (BC, Canada) and Alberta Newsprint Company (AB, Canada), respectively. Hemlock was second-stage outlet TMP refined at 1366 kWh/t, while spruce/pine pulp was taken from a primary stage HC refiner set at 1410 kWh/t. The obtained pulp was homogenized by a Hobart Mixer and stored at 4°C until used for further experimentation.

Chelation

Each of the pulps (60 g oven-dry) was chelated with 0.6% diethylene triamine pentaacetic acid (DTPA) at 4% consistency and 60 °C for 30 min. Following chelation, pulps were washed with deionized water and dewatered to about 20% consistency. All pulps used for our investigation were chelated through this method prior to experimentation.

Preparation of Chlorine Dioxide

The associated reaction for chlorine dioxide generation is shown below.

$$NaClO_3 + \frac{1}{2}H_2C_2O_4 \cdot H_2O + H_2SO_4 \rightarrow ClO_2 + CO_2 + NaHSO_4$$

A reaction flask was immersed in a water bath heated between 60 and 70°C. Sodium chlorate and oxalic acid, which acted as a reducing agent, were added to the flask, followed by sulfuric acid solution, which was used to catalyze the reaction. The resulting gas mixture was scrubbed in a trap flask that held deionized water at room temperature before reaching a cooled four-liter amber receiving bottle that was placed in an ice bath. The amber bottle contained 2 L of deionized water to dissolve the generated chlorine dioxide. Finally, any remaining chlorine dioxide or acids within the gas mixture were scrubbed in a second trap flask containing 1M sodium hydroxide.

Chlorine Dioxide Treatment

Prior to treatment, the concentration of the chlorine dioxide solution was measured by addition of potassium iodide followed by back-titration of the released iodine with sodium thiosulphate. Chelated pulps (60 g oven-dry) were placed into sealable polyester bags and mixed with chlorine dioxide solution at 10% treatment consistency before incubating in a water bath that was set between 25-65°C for 10-120 min. The pulp was mixed every 5 min. The treated pulp was removed from the bath and

45 mL of its filtrate was extracted for later pH and COD measurements. The pulp was washed using deionized water and the filtrate was recycled to retain fines.

Second Stage Alkaline Soaking/Alkaline Peroxide Treatment

Treated pulp (30 g oven-dry) was placed into a one-liter flask and the appropriate quantity of sodium hydroxide was dissolved into deionized water. The pulp was then mixed with the sodium hydroxide solution at 4% treatment consistency. The resulting pulp suspension was left to soak at room temperature for one hour. After completion of the soak, the filtrate was collected for later pH and COD measurement. The pulp was washed with deionized water and the filtrate was recycled to retain the fines.

Alternatively, some of the pulps were treated instead with alkaline peroxide. Chlorine dioxide treated pulp (30 g oven-dry) was placed into a plastic bag and mixed with bleaching liquor at 15% treatment consistency, where the bleaching liquor consisted of 4% w/w sodium hydroxide, 4% w/w hydrogen peroxide, 0.1% w/w magnesium sulfate and 3% w/w sodium silicate. The pulp was immersed in a water bath for 70 °C for 90 min and mixed every 10 min. Afterwards, the pulp was washed with deionized water and the filtrate was recycled to retain fines.

Refining

Low consistency refining was simulated through refining with a "Waring Blendor", which consists of a one-liter capacity bowl and blades. The pulp was blended as a 500 mL suspension at 2.4% consistency for 10, 20, or 30 min. The "Waring Blendor" was operated at low speed, with an input voltage and current of 115 V and 1.9 A respectively (11).

Acid group content

The amounts of carboxylic and sulfonic acid groups within the pulp were measured using the method of conductometric titration described by Beatson (12).

Pulp yield measurement

The pulp yield was measured through chemical oxygen demand, using a relationship between yield and COD described by Genco (10). The COD in pulp filtrates from the treatments was determined using the standard USEPA Reaction Digester Method 8000. Briefly the filtrates were reacted with potassium dichromate in 50% sulfuric acid solution at a temperature of 150 °C to oxidize organic matter within the filtrate and the consumed potassium dichromate was measured colormetrically.

Pulp freeness measurement

Pulp freeness was determined using PAPTAC standard method C.1. Before freeness measurement and handsheet making, pulps were hot disintegrated to remove latency within the suspension.

Handsheet Properties

All handsheets were prepared following PAPTAC method C.4. The white water was recycled to retain fines during handsheet making. Brightness, bulk, tear resistance and tensile strength were determined using PAPTAC methods D.12, D.4, D.9, and D.6H respectively. The light scattering coefficient was measured following TAPPI method T1214.

RESULTS AND DISCUSSION

Chlorine dioxide is commonly used to oxidize lignin in the kraft pulp bleaching process and the reactions mechanisms have been well explored. As shown in Figure 1, the main reaction is electrophilic attack at phenolic lignin units to form the radical which undergoes oxidation to produce an intermediate which decomposes to either a muconic acid or a quinone (6). The formation of carboxylic acid groups provides the potential for pulp strength increase, especially if converted to the sodium form, whereas quinone formation will lead to colour generation.

The pulp strength increase could be exploited by reducing the refining energy input required to achieve a given tensile strength or possibly for the production of mechanical pulps with superior strength required for replacement of kraft pulp in higher grade products such as high-strength packaging paper. In order to explore these possibilities two softwood thermomechanical pulps of high freeness were chosen for treatment with chlorine dioxide followed by a soaking in sodium hydroxide solution, and the pulp properties were evaluated. A high-strength treated pulp was further refined on a laboratory-scale at low consistency to determine the full potential of using this process to produce a low energy mechanical pulp capable of substituting for unbleached kraft in packaging papers.

Chlorine Dioxide Treatment of Hemlock Thermomechanical Pulp

A thermomechanical hemlock pulp refined at specific energy of 1366 kWh/ADT to 380 mL freeness was obtained from Catalyst Paper (BC Canada). The pulp was treated with chlorine dioxide over a range of charges, times and temperatures. The chlorine dioxide treatments were followed by soaking in sodium hydroxide solution or with an alkaline peroxide bleach. Pulp strength and optical properties were assessed.

Strength Properties

Figure 2 shows the effect of chlorine dioxide charge and treatment time on tensile strength of the treated pulp. As shown in Figure 2a, for treatments of 45 min at 45 °C and 10% consistency, a 3 N·m/g increase in tensile strength was observed when 1% w/w chlorine dioxide was applied. There was little further increase with increasing charge to 4.5% w/w chlorine dioxide. When treating the pulp with 3% w/w chlorine dioxide at 45 °C and 10% consistency, there was only a small increase in tensile strength by increasing the treatment time from 10 to 90 min (Figure 2b). Obviously, there is a limit to the tensile strength that can be reached. However, for treated pulps, an alkali soak (0.4% w/w NaOH, 23 °C, 4% consistency, 60 min) increased the tensile by 10 N·m/g to around 43 N·m/g (Figure 2b). It should be noted that both chlorine dioxide treatment and alkali soaking are essential for the tensile increase, as evidenced (Figure 2b) by the lack of any tensile gain for the original TMP on soaking in sodium hydroxide solution.

Figure 3 shows the results of replacement of the alkali soak with an alkaline peroxide bleach of the pulp treated with 1% w/w chlorine dioxide at 45 °C, 10% consistency for 10-120 min. As shown, alkaline peroxide bleach resulted in tensile index values of around 43 N·m/g. However, in this instance, there was also significant gain in the tensile when alkaline peroxide bleach was performed on the original pulp. This is consistent with our earlier work on improving TMP tensile strength through the application of highly alkaline hydrogen peroxide (11).

It is interesting to postulate why increased pH is required to activate the effects of the chlorine dioxide treatment. Chlorine dioxide treatment oxidises lignin, generating carboxylic acid groups in the pulp. The carboxylic acid group content of the original hemlock TMP doubled on treatment with chlorine dioxide from 0.06 to 0.12 mmol/g OD pulp. As shown in Figure 4, during chlorine dioxide treatment, the pH of the pulp suspension drops, mainly from the formation of hydrochloric acid. At pH values below 2, the carboxylic acids generated in the fibres and fines will be in their non-dissociated form, and conversion to the ionised sodium form by soaking in sodium hydroxide solution is necessary to produce swelling and softening of the fibres and fines (13). It is this swelling and softening that results in increased sheet consolidation and bonding, which in turn leads to increased tensile strength. We observed similar tensile increase on alkali soaking for treatment of TMP with ozone which is done under mildly acidic conditions (1). The tensile gain after treatment with 5% ozone was 12 N·m/g but increased to 33 N·m/g after an alkali soak.

The tear indices of the pulps were around 10 mN·m^2/g with a drop to around 8 mN·m^2/g at the higher tensile levels (data not shown).

Optical Properties

Chlorine dioxide treatment lowered the hemlock TMP brightness by up to 18 units, the decrease being more pronounced at the higher chlorine dioxide charges (Figure 5a). Moreover, the alkali soak of chlorine dioxide treated pulp further decreased the pulp brightness. For example, regarding the treatment with 1% w/w chlorine dioxide at 45 °C and 10% consistency, the alkali soak further decreased the brightness from 42.5% ISO to 35% ISO (Figure 5b). The brightness decrease on chlorine dioxide treatment can be attributed to the formation of quinones which is a known product from the reaction of chlorine dioxide with lignin (6). Indeed, the formation of a pink coloration was evident on adding the chlorine dioxide to the pulp. The further loss in brightness on alkali soaking maybe related to the extractives in hemlock. This brightness drop wasn't found in our work with the pine/spruce furnish (see later).

To recover the brightness of the dark chlorine dioxide treated pulps, alkaline hydrogen peroxide treatment was shown to be an effective strategy. As shown in Figure 5c, the dark pulps, obtained from treatment with 1% w/w chlorine dioxide at 45 °C and 10% consistency, were readily brightened by treating with 4% hydrogen peroxide at 4% NaOH and 70 °C for 90 min.

Indeed, the chlorine dioxide treated pulps responded to brightening better than the original pulp, achieving brightness levels of 71.5% ISO compared to 67.5% ISO.

Chlorine Dioxide Treatment of Pine/Spruce Thermomechanical Pulp

To assess the full potential of chlorine dioxide treatment in strengthening TMP is was of interest to apply the process to a higher freeness pulp. A thermomechanical pulp produced from a mix of spruce/pine chips refined with a specific energy of 1410 kWh/t was obtained from Alberta Newsprint Company (Alberta, Canada). In a similar manner to the hemlock pulp, the pine/spruce TMP was treated with chlorine dioxide over two charges (1 and 5%), times (10-60 min) and temperatures (45-55 °C). The chlorine dioxide treatments were followed by an alkali soak over a range of sodium hydroxide charges (0.4-3% w/w). A treated pulp with high strength was chosen for further refining at low consistency on a laboratory scale. Pulp strength and optical properties were assessed.

Strength Properties

As shown in Figure 6a, under the investigated conditions, the tensile strength of the treated pulp only slightly increased by increasing the chlorine dioxide charge from 1% to 5% w/w. When the chlorine dioxide treated pulps were soaked with sodium hydroxide for 60 min, increases in pulp tensile became evident (Figure 6b). The increase in tensile on soaking with 1% sodium hydroxide for 60 min was 28% to around 25.5 N·m/g. The increase was similar for both 1% and 5% chlorine dioxide treatments. With increasing the charge of sodium hydroxide used for soaking, a further increase in tensile strength for the TMP treated at the higher chlorine dioxide charge (5% w/w) was observed. For example, at a sodium hydroxide charge of 3% w/w, a tensile strength of 32 N·m/g was achieved. However, no further increase in tensile strength was obtained by soaking the 1% w/w chlorine dioxide treated samples with increasing sodium hydroxide charge (Figure 6b).

A likely reason might be that lower amounts of acid groups are generated at the lower charge of chlorine dioxide and 1% w/w sodium hydroxide is sufficient to neutralize all of them under the conditions used for the soak. In addition, it should be noted that no tensile gain was observed on soaking the untreated pulp with 1% w/w sodium hydroxide charge. These observations are similar to those found with the lower freeness hemlock TMP, in which the benefits of the chlorine dioxide treatments are not realised until a subsequent alkali soak.

As postulated previously, the strengthening effect is probably created by the formation of sodium salts of the carboxylic acids which were generated from the oxidation of lignin by chlorine dioxide. These sodium salts then lead to softening and swelling of the fibres and fines and resultant increased bonding and sheet consolidation, which is consistent with the negative the correlation between tensile strength and bulk of the paper (Figure 7a). Also, the decreased pulp freeness can be another evidence for the proposed mechanism (Figure 7b).

Development of Strength Properties on Low Consistency Refining

The spruce/pine TMP before and after treatment with chlorine dioxide (1% chlorine dioxide, 55 °C, 10 min, 10% consistency) plus alkali soak (1% NaOH, 60 min, 23 °C, 4% consistency) were refined in a "Waring Blendor" at 2.4 % consistency for 10, 20 and 30 min. This procedure has been previously shown to mimic pilot-scale low consistency refining with each 10 min of blending corresponding to the application of a specific energy of about 110 kWh/t (11).

The expected drop in freeness on refining was observed for both pulps (Figure 8a). The treated pulp started with a lower freeness than the control non-treated pulp reflecting more consolidation of the pad during the freeness measurement, but the rate of freeness drop on refining was similar for both pulps. Figure 8b shows the effect of refining on the improvement of the tensile strength for the two pulps. As shown, the initial tensile strength of the treated pulp was higher than the control, consistent with reported above, and increased by 15.6 N·m/g after refining for 30 min, to final tensile of 41 N·m/g. The tensile gain for the control pulp on refining was less, at 11.8 N·m/g, to reach a final tensile of 30.6 N·m/g. Apparently, not only did the chlorine dioxide treatment followed by the alkali soak increase the initial tensile strength, the treatment produced a pulp that was easier to refine. To achieve a tensile of 30 N·m/g took 23 min less refining for the treated pulp which is equivalent to a reduction in refining energy of around 253 kWh/t. Judging by the trend lines, the energy savings would be even greater at higher tensile values. For both pulps, the tear strength increased with tensile as seen in Figure 8c. The optical properties of these pulps are discussed in the following section.

Optical Properties

As was seen with the Hemlock TMP, chlorine dioxide treatment lowered the brightness of the pine/spruce pulp (Figure 8). This brightness drop was more severe for the higher chlorine dioxide charge. In contrast with the Hemlock TMP, the TMP from the pine/spruce furnish did not exhibit a further brightness drop on alkali soaking. This maybe related to the nature of the extractives in the different species.

Effect of Low Consistency Refining on Optical Properties

After the initial drop in brightness on chlorine dioxide treatment followed by an alkali soak (for conditions see above), refining increased pulp brightness through the generation of fines and increased scattering coefficient (Figure 9). The decrease in scattering coefficient on chemical treatment is consistent with increased bonding between the fibres and fines and improved tensile strength.

CONCLUSIONS

Chlorine dioxide treatment followed by alkali soak or alkaline peroxide bleach was demonstrated to be an effective strategy to improve the strength properties of the high freeness thermomechanical pulps. In addition, the loss in pulp brightness caused by chlorine dioxide treatment could be readily recovered by alkaline peroxide bleaching, achieving higher brightness than the bleached original pulp. The chlorine dioxide-alkali treated pulp responded better to low consistency refining, with less energy being required to achieve a given tensile strength. Low consistency refining improved the pulp brightness and scattering coefficient.

REFERENCES

1. Chang XF, Olson JA, Beatson RP. A comparison between the effects of ozone and alkaline peroxide treatments on TMP properties and subsequent low consistency refining. BioResources. 2011;7(1):99.

2. Chang XF, Miller M, Sun Y, Olson J, Beatson R. A pilot scale comparison of the effects of chemical treatments of wood chips on the properties of high freeness TMP. In the proceedings of the International Mechanical Pulping Conference, September 26-28, 2016, Jacksonville, Florida, USA.

3. Chang XF, Luukkonen A, Olson J, Beatson R. Pilot-Scale Investigation into the Effects of Alkaline Peroxide Pre-Treatments on Low-Consistency Refining of Primary Refined Softwood TMP. BioResources. 2016;11(1).

4. Yu S, Miller M, Chang XF, Olson J, Beatson R. A pilot-scale comparison of the effects of chemical pre-treatments of wood chips on the properties of low consistency refined TMP. In the proceedings of the International Mechanical Pulping Conference, September 26-28, 2016, Jacksonville, Florida, USA.

5. Dence CW, Reeve DW. Chapter IV: Section 3: Chlorine Dioxide in Delignification. In: Dence Carlton, Reeve D, editors, Pulp Bleaching; Principles and Practice. Publ. TAPPI. USA, 1996. p. 263-290

6. Sjostrom E. Wood chemistry: fundamentals and applications. 2nd ed. Academic Press: San Diego, Calif; 1993.

7. Dence C. Section III: Chapter 3: Chemistry of Chemical Pulp Bleaching. In: Dence Carlton, Reeve D, editors. Pulp Bleaching: Principles and Practice. Publ. TAPPI USA; 1996. p. 127-153.

8. McDonough TJ, Andrews D.H. Chemimechanical pulps from mixed Northeastern hardwoods by oxidation and alkaline treatment. Pulp and Paper Canada. 1980 June;6(2):43.

9. McDonough T. Simultaneous brightening and strengthening of GW, TMP and high-yield bisulphite pulp. Pulp Paper Can. 1977 December,78(12):T290.

10. Genco JM, Zorn KD, Cole B. Estimation of pulp yield loss in oxygen delignification using COD measurements. TAPPI Pulping Conference, Boston, MA; 2000.

11. Chang XF, Bridges C, Vu D, Kuan D, Kuang L, Olson JA, et al. Saving electrical energy by alkaline peroxide treatment of TMP prior to low consistency refining. J Pulp Paper Sci. 2010;36(3-4):129-34.

12. Beatson RP. Determination of sulfonate groups and total sulfur. In: Methods in lignin chemistry. Springer; 1992. p. 473-84.

13. Scallan AM, Grignon J. The effect of cations on pulp and paper properties. Svensk Papperstidning. 1979;2:40-47.

ACKNOWLEDGEMENTS

This work was funded by the Natural Sciences and Engineering Research Council of Canada through the Collaborative Research and Development program and through the support of our partners: AB Enzymes, Advanced Fibre Technologies, Alberta Newsprint Company, Andritz, BC Hydro, Catalyst Paper, Canfor, FPInnovations, Holmen Paper, Meadow Lake Pulp (Paper Excellence), Millar Western, NORPAC, West Fraser (Quesnel River Pulp and Slave Lake Pulp), Westcan Engineering, and Winstone Pulp International. Special thanks to Catalyst Paper and Alberta Newsprint Company for supplying the pulps.

FIGURES

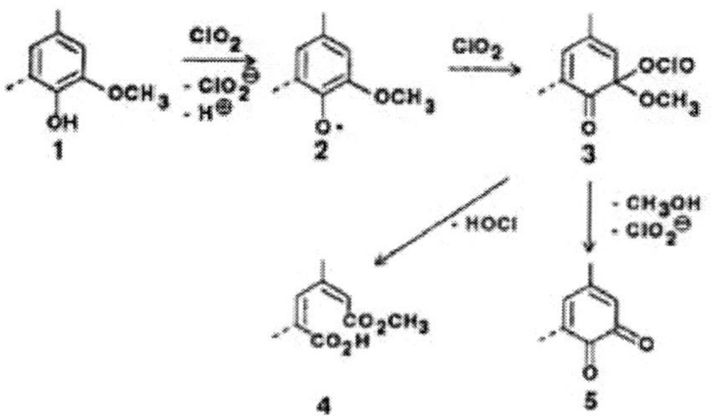

Figure 1: Reaction of chlorine dioxide with lignin to form muconic acids (4) and quinones (5): From Sjostrom (6)

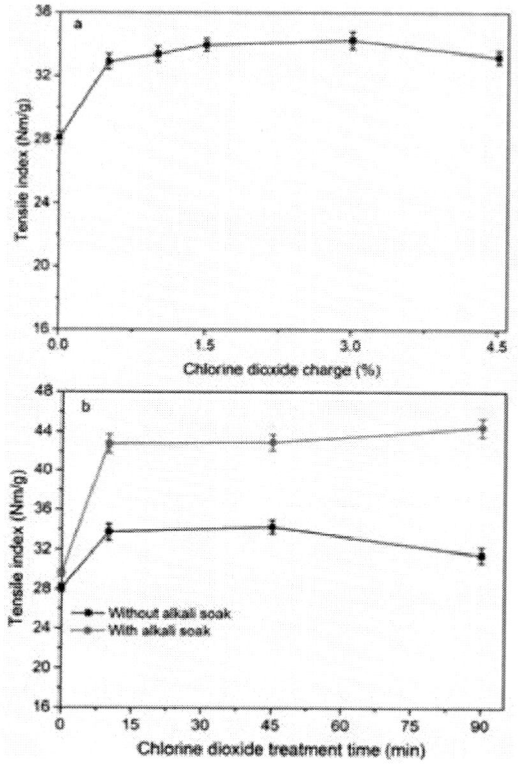

Figure 2: Effect of chlorine dioxide charge and treatment time on tensile strength of hemlock TMP. Treatment conditions 45°C and 10% consistency. (a): Treatment time; 10 min. (b) 3% w/w chlorine dioxide charge.

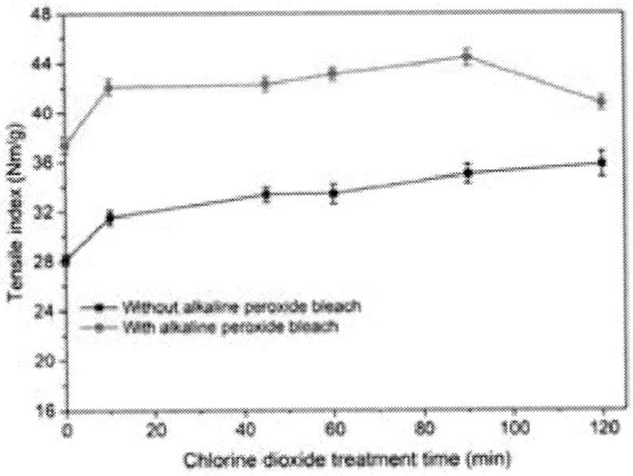

Figure 3: Effect of alkaline peroxide leaching on the tensile strength of chlorine dioxide treated pulp. Chlorine dioxide treatment conditions: 1% charge, 45°C, 10% consistency.

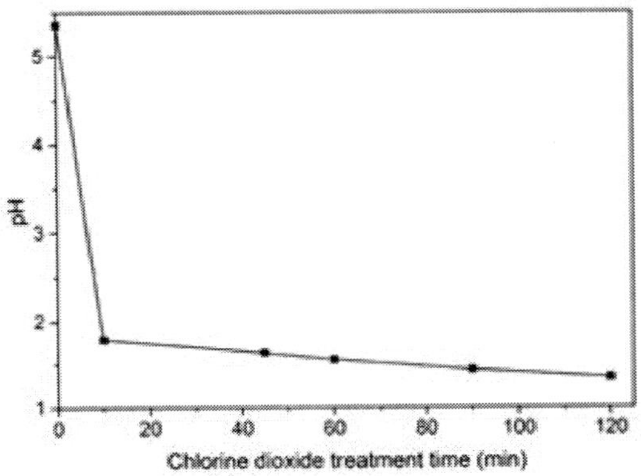

Figure 4: effect of chlorine dioxide charge on the pH of the pulp suspension. Treatment conditions: chlorine dioxide charge 1%, 45°C, 10% consistency

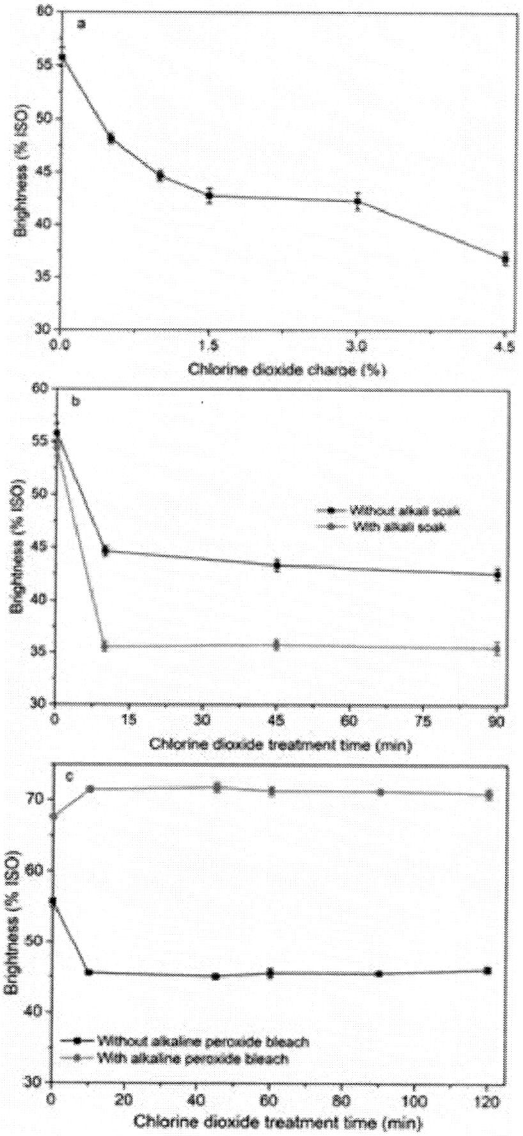

Figure 5: Effect of chlorine dioxide treatment, alkali soak and alkaline peroxide bleach on pulp brightness. Chlorine dioxide treatment was carried out using 1% chlorine dioxide at 45 °C and 10% treatment consistency, Effect on brightness of (a) chlorine dioxide charge (b) chlorine dioxide treatment time with and without alkali soak, (c) chlorine dioxide treatment time with and without alkaline peroxide bleaching.

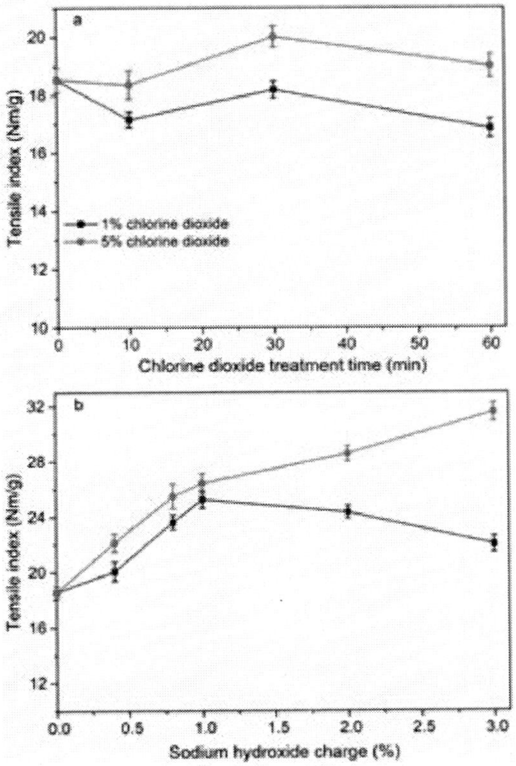

Figure 6: Effect of chlorine dioxide treatment and alkali soak on the tensile strength of the spruce/pine pulp. Chlorine dioxide treatment was carried out with 1% and 5% chlorine dioxide at 55°C and 10% consistency. (a) effect of chlorine dioxide treatment time, (b) effect of sodium hydroxide charge.

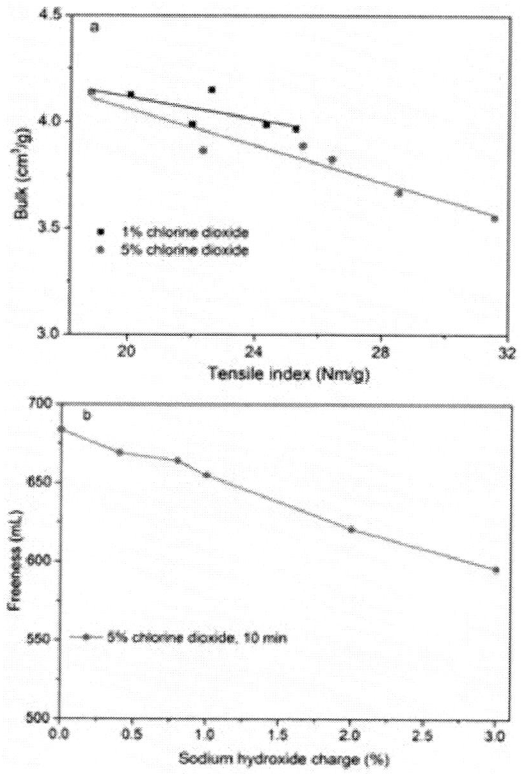

Figure 7: Effect of chlorine dioxide treatment on the spruce/pine TMP properties. Chlorine dioxide treatments were carried out at 55°C and 10% treatment consistency (a) correlation between bulk and tensile strength at different chlorine dioxide treatment times, (b) pulp freeness at different sodium hydroxide charges.

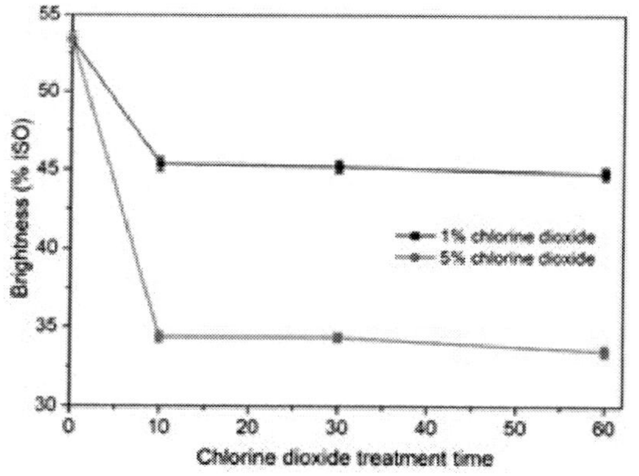

Figure 8: Effect of chlorine dioxide treatment time on the spruce/pine brightness. Treatments were carried out at 1% and 5% chlorine dioxide charge, 45°C and 10% treatment consistency.

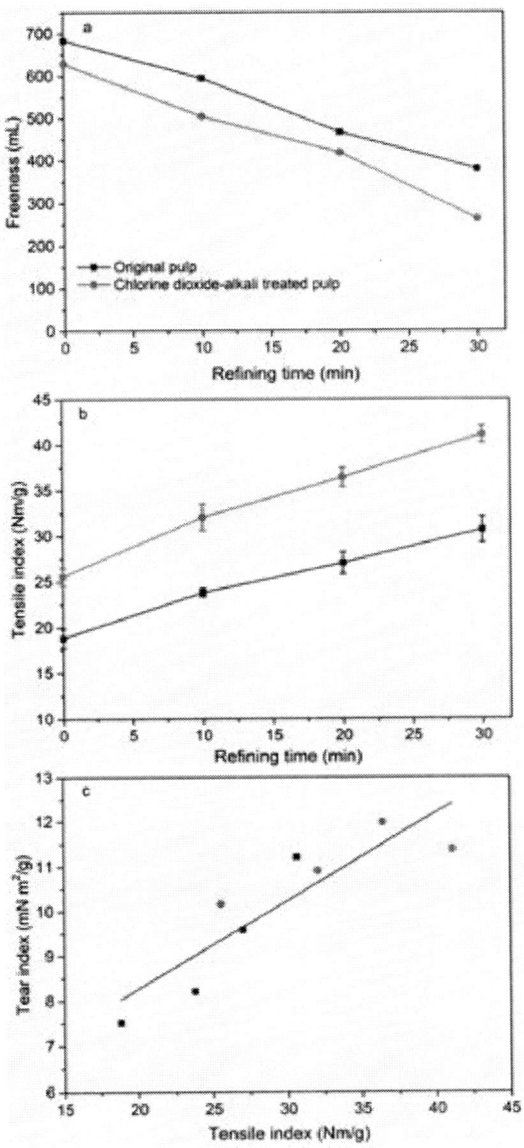

Figure 9: Effect of low consistency refining on the properties of chlorine dioxide treated spruce/pine TMP. Chlorine dioxide treatment was carried out at 1% chlorine dioxide charge for 10 min at 45°C and 10% treatment consistency. The following alkali soak was carried out at 1% sodium hydroxide charge for 60 min at 23°C and 4% treatment consistency. (a) pulp freeness, (b) tensile index, (c) correlation between tear index and tensile index.

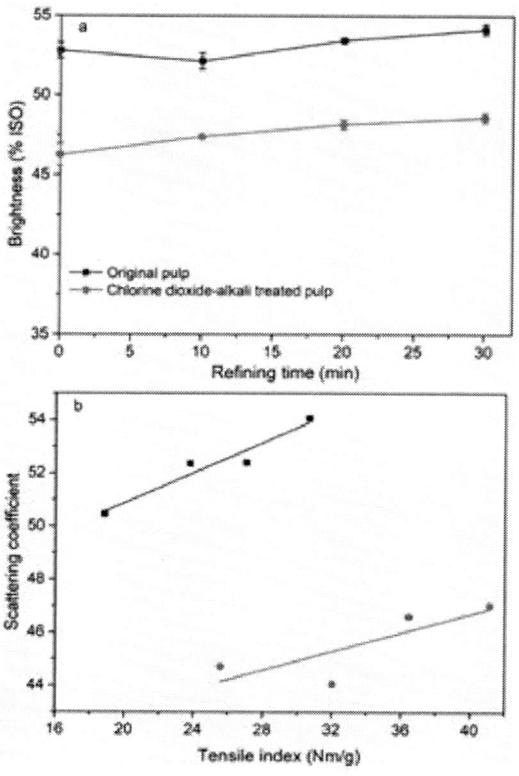

Figure 10: Effect of low consistency refining on the optical properties of spruce/pine TMP. Chlorine dioxide treatment was carried out at 1% chlorine dioxide charge for 10 min at 45°C and 10% treatment consistency. The following alkali soak was carried out at 1% sodium hydroxide charge for 60 min at 23°C and 4% treatment consistency. (a) brightness, (b) correlation between scattering coefficient and tensile index.

538939 Understandings of Thermal Transformation of Cellulose Surface and Crystalline Core By *in-Situ* Nonlinear Vibrational Spectroscopy

Abstract:

In this study, Total Internal Reflection Sum Frequency Generation Vibrational Spectroscopy (TIR-SFG-VS) combined with non-TIR-SFG-VS enabled detection of the molecular structures of surface layers and the crystalline core of cellulose, which correlates with the biomass recalcitrance. From the SFG spectra in the C-H and O-H regions, it was found for the first time that the surface layers of IÎ^2 cellulose differed in structural and spectroscopic signatures with its crystalline core. This work demonstrates the capacity of TIR and Non-TIR SFG-VS in selectively studying the structures and polymorphs of cellulose. This discovery is significant because it not only challenges the traditional understandings of cellulose materials that neglects the differences between its surface and core region as well as their effects on the cellulose deconstruction reactivity, but also provides a novel spectroscopic tool to measure the spectral signature of the bulk crystalline core and the surface layer of cellulose. The implementation of aqueous pretreatment of cellulose using a heated fluid test bed with dynamic TIR-SFG-VS analysis revealed the structural changes of Avicel and cellulose IÎ^2 caused by thermal effects. The different crystalline structures of Avicel and cellulose IÎ^2 changed during heating and cooling processes. The recrystallization to a new crystalline structure in the process of cooling could strongly affect the recalcitrance of cellulose, leading to new insights on the effectiveness of biomass pretreatment.

520615 Fabrication and Characterization of Novel Cellulose Acetate Hollow Fiber Nanoporous Membranes Prepared Via Thermally Induced Phase Separation

Abstract:

Cellulose acetate (CA) is a kind of environment-friendly and biodegradable material and CA separation membrane has been widely used in the fields of food processing, pharmaceutical engineering, medicine, etc.. Hydrophilic modified cellulose acetate (CA) hollow fiber nanoporous membrane was firstly fabricated via thermally induced phase separation (TIPS) method. The hydrophilic additive was added into the dope solution. The effects of additive on membrane morphologies, hydrophilicity, permselectivity of CA membrane were investigated systematically. With the increase of additive content, the fabricated membrane had the denser cross-section pore structure and the thicker surface layer. The morphological change was beneficial to the membrane ductibility, corresponding to the increase of elongation from 20.0 % to 74.8 %. The additive played a critical role in the improvement of membrane hydrophilicity. The initial contact angle sharply decreased from 79° of CHM25 membrane to 53° of CPHM25-4 membrane. The additive molecules presented excellent twine with CA molecules which facilitated the improvement of membrane performance stability. The preferred CPHM25-2 membrane had the ultrafiltration membrane characteristics with high rejections for different neutral solutes. The molecular weight cut-off (MWCO) was about 13000 Da and the pure water flux increased 8 times compared with CHM25 membrane. Furthermore, heat treatment were adopted to compact the surface pore size in order to realize the membrane separation performance transition from ultrafiltration to nanofiltration. As for CPHM25-2-80-50 membrane, the rejections of Na_2SO_4 and NaCl solution could reach 89.6 % and 29.0 %. It is promising that the novel hydrophilic modified CA hollow fiber nanoporous membrane has broad application prospects in the fields of food processing, chlor-alkali industry, wastewater treatment, etc..

538805 Hairy Cellulose Nanocrystals-Colloidal Starch Nanocomposite Coatings with Nanoengineered Viscosity Improve the Mechanical Properties of Papers: One Stone, Two Birds

Abstract:

Emerging paper-based technologies encompassing lab-on-a-chip platforms, microfluidics, sensors, actuators, flexible electronics, and energy-storage devices, such as batteries, demand mechanically robust substrates. Current gold standards for paper and paper board reinforcement rely on petroleum-based binders and coatings, such as latex beads cured with melamine formaldehyde and urea formaldehyde resins, which have raised the red flag because of their environmental footprints. Here, we introduce an all-green reinforcing nanocomposite based on the ammonium zirconium carbonate (AZC)-mediated synergistic crosslinking of starch nanoparticles (biolatex) and a newly emerged type of nanocelluloses named anionic hairy cellulose nanocrystals. AZC, an environmentally friendly and cost-effective crosslinker, reacts with the hydroxyl groups of starch and cellulose nanocrystals, while the carboxylic acid groups of hairy nanocelluloses maintain the solution viscosity low, favoring industrial transportation, which may not be possible using conventional (nonhairy) nanocelluloses. Synergistic action of these green nanomaterials improved the mechanical properties of a glass paper by 1 order of magnitude within a broad range of temperature (T 25–75 °C) and relative humidity (RH 0–80%) at noticeably low nanoparticle concentrations (≤1.5 wt %) in the reinforcing dispersions. This technology may set the stage for the next generation high-performance green, renewable, and biodegradable reinforcing coatings and nanocomposites.

540773 2018 Outlook for Energy: A View to 2040

Abstract:

The Outlook for Energy is ExxonMobil's global view of energy demand and supply through 2040. It also highlights the dual challenge of providing affordable energy to support prosperity while reducing environmental impacts – including the risks of climate change. We use the Outlook to help inform our long-term business strategies and investment plans.
The Outlook anticipates significant changes through 2040 across the world to boost living standards, reshape the use of energy, broaden access to abundant energy supplies, and accelerate decarbonization of the world's energy system to address the risks of climate change.

- Global energy needs will rise about 25 percent over the period to 2040
- Electricity demand will nearly double in non-OECD countries
- Worldwide electricity from solar and wind will increase about 400 percent
- Natural gas will expand its role, led by growth for electricity generation
- Growth in oil demand will be driven by commercial transportation and chemical needs
- Efficiency gains and growing use of less-carbon intensive energy sources will contribute to a nearly 45 percent decline in the carbon intensity of global GDP

Energy is fundamental to modern life, and as the world's population approaches 9 billion people in 2040, we are challenged to improve living standards everywhere. Understanding the factors that drive the world's energy needs – and likely choices to meet those needs – is the mission of the Outlook. By sharing the Outlook with the public, we hope to broaden that understanding. Energy matters to everyone, and we all play a role in shaping its future.

541543 Energy Decarbonisation Scenarios

Abstract:

With the energy sector being associated with 2/3 of the annual greenhouse gas emissions, decarbonizing the energy is a key step to reduce their level. The National Determined Contributions (NDC's) announced so far by the participating countries do not meet the objectives of the Paris-climate agreement, as far as reaching a peak in emissions in the near future or decreasing to a level that is compatible with the stated long-term targets.

Using different scenarios for energy demand and supply, options for energy sector decarbonisation are evaluated. Meeting the 2-degree target requires a major transformation in the power sector, efficiency improvements – specially in the buildings and in the industry, and a revolution in transport towards fuel and modal shifts. With a significant increase in renewable energy, system integration and optimization will also be key in reducing the cost of the energy transition. Meeting more ambitious climate goals ("well below 2 degrees") will push further the technological innovation and implementation requirements throughout the entire energy value chain. A scenario aiming at zero net-emissions by 2060 is presented, along with a comparison to the 2-degree case.

542625 Fundamental Research Needs to Advance Energy Technologies

Abstract:

The U. S. Department of Energy (DOE), Office of Science, Office of Basic Energy Sciences (BES) supports fundamental research in chemical and materials sciences to provide the foundations for new energy technologies and to support DOE missions in energy, environment, and national security. This presentation will discuss opportunities for fundamental research to impact DOE's energy mission "to catalyze the timely, material, and efficient transformation of the nation's energy system and secure U.S. leadership in energy technologies" with a focus on the way we generate, store and use energy nationally. I will provide an overview of BES strategic planning over the past decade that identified priority research directions for advancing energy applications, highlight key scientific advances in these areas, and discuss some future opportunities for modern science, particularly chemical sciences, to accelerate the transformation of the U. S. energy portfolio.

528883 Towards Economical and Sustainable Production of Wood Based Nanomaterials

Abstract:

In this presentation, I will outline the current status of cellulose nanomaterial production including the deficiencies of existing technologies. Then, I would like to discuss some of our research on integrated production of cellulose nanocrystals (CNC) and fibrils (CNF) with full recovery of chemicals. Specifically I will present our recent study using dicarboxylic acids (DCAs) for sustainable and potentially economical production of highly thermal stable and functional (carboxylated) cellulose nanomaterials. DCA hydrolysis overcomes several disadvantages of existing technologies (mineral acid or enzymatic hydrolysis, oxidation such as TEMPO-mediated), such as (1) low thermal stability which prevented their applications in the largest market â€“ wood based composites processed at elevated temperatures such as extrusion; (2) difficulties in economic recovery of chemicals. DCAs can be easily and economically recovered through commercially proven crystallization process at ambient temperatures. DCA hydrolysis (DCAH) resulted in low CNC yield, but minimal cellulose loss or high solids yield. The remaining partially hydrolyzed fibers can be subsequently fibrillated into CNF to realize integrated production of CNC with CNF with tunable yield, properties, and morphologies. The DCAH-CNC has carboxylic acid group content of 0.1 â€“ 1.5 mmol/g CNC. The DCAH-CNF was also functionalized with surface carboxyl groups for dispersion. The low strength of DCA also resulted in CNC with higher crystallinity and longer lengths of 300 â€“ 600 nm than those produced using mineral acids, suitable for polymer reinforcement.

534580 High Shear Capillary Rheometry of Cellulose Nanomaterials for Industrial Relevant Processing

Abstract:

Nanocellulosics, which are typically found in the form of Cellulose nanocrystals (CNCs) and Cellulose Nanofibrils (CNFs) provide promise as both filler and reinforcement materials. The biological origin of CNCs promises a biorenewable resource combined with potential expedited degradation times compared to traditional plastic species. Meanwhile, the surface functional groups on CNCs promise networking capabilities via hydrogen bonding. While much research is currently investigating the possible uses of this material, they offer limited aid if they are not scalable to industrial processing techniques. Common processing techniques such as injection molding subject materials to strain rates upwards of 10,000 s^{-1}. The resulting stresses have the potential to cause alignment in the crystals, providing exploitable anisotropy in any injected product. In this work, we investigate the effect of these strain rates and subsequent shear stresses on CNCs and CNFs over multiple concentrations up to 12.1 wt% in water. Both capillary and extensional rheometry have been deployed to characterize the potential degradation and relaxation characteristics of these materials after exposure to high strain rates. Furthermore, we investigate the potential to apply Williams-Landel-Ferry (WLF) like shift factors for time-temperature superposition as well as time-concentration superposition.

535197 Towards Standardization of Laboratory Preparation Procedure for Uniform Cellulose Nanopapers

Abstract:

Towards Standardization of Laboratory Preparation Procedure for Uniform Cellulose Nanopapers

Cellulose nanopapers, a 2D flat and foldable film made of network-forming cellulose nanofibers (CNF), are strong, tough and biodegradable materials of great interest as they have potential to replace the conventional fossil based polymers in various applications. Although there are several cellulose nanopaper preparation methods described in literature, no standard method is available in order to prepare uniform and smooth nanopapers within a practical time frame. In present study, TAPPI standard method of preparation of pulp handsheets was systemically modified/optimized to prepare flat and wrinkle free cellulose nanopapers in few hours of preparation time. Physical properties of these cellulose nanopapers such as tensile strength and water vapor transmission rate were comparable to the reported values in literature. Proposed standard preparation method will enable the comparison of the physical properties of cellulose nanopapers prepared in different studies.

Mechanical and barrier properties of optimized cellulose nanopapers (standard deviations are indicated in bracket)

Sample	Burst Index (kN/g)	Folding Endurance	Tear Index ($mN.m^2/g$)	WVTR ($g/m^2.day$)	Tensile Index (kN/kg)	Drainage Time (minutes)
CNF (2 Liter suspension)	9.95 (0.7)	3.6 (0.16)	2.87 (0.08)	115 (3.8)	93.3 (8.0)	12 (0.6)

538125 Emerging Cellulose Nanocrystals for Threshold Scale Inhibition: A Step Forward in Universal Biomass-Based Crystal Engineering

Abstract:

Scales, sparingly soluble inorganic salts, are formed in aqueous systems with high levels of hardness and alkalinity. The precipitation and adsorption of scale in industrial unites have turned into a major issue, imposing considerable operational and economic losses to industries. The current gold standard in scale prevention involves phosphonated macromolecules, which have raised a red flag due to severe anthropogenic side effects, such as acidification and eutrophication. Here, we provide a conceptual solution to engineer the building blocks of plant cell walls, cellulose fibrils, at nanoscale, to overcome the structural and chemical limitations of conventional nanocelluloses. We have developed the first family of threshold cellulose-based scale inhibitors with promising capabilities in inhibiting and modifying calcium carbonate, the most common type of scale. Cellulose fibrils are disintegrated into dicarboxylic acid functionalized biopolymers and hairy nanocelluloses, which, at ppm concentrations, are able to completely prevent the calcium carbonate crystallization under electrochemically stimulated harsh scaling conditions, where an additivefree system fully scales in less than 0.5 h. The nanostructure and chemistry of cellulose nanocrystals are investigated to shed light on the mechanism of scale inhibition, detailing the inevitable structural limitation of conventional nanocelluloses and the promising role of dicarboxylated cellulose. The outcome of this research may translate into a universal biomass-based sustainable, green solution for the (i) long-lasting scaling challenge of water-based industries, (ii) efficient design of inorganicâ€“organic biomimetic nanocomposites, and (iii) improved solubility of sparingly soluble species based on the most abundant biopolymer in the world.

531924 Solid-State Depolymerization and Isolation of Lignin from Lignocellulosic Biomass

Abstract:

This study demonstrated that lignin could be efficiently depolymerized in the solid state, with minimal condensation, and separated from dissolved and hydrolyzed cellulose and hemicelluloses, in an inorganic ionic liquid (acidic lithium bromide trihydrate (ALBTH) system) under mild conditions (with 40 mM HCl at 110 °C). The resultant ALBTH lignin contained abundant uncondensed moieties (i.e., Hibbert's ketones and benzodioxanes), from partial depolymerization, and could readily be isolated from biomass. Reactions using lignin model compounds confirmed the formation of the uncondensed moieties and revealed the synergy between LiBr and acid in inducing the crucial intermediate α-benzyl carbocations, which then led to cleavage of the β–*O*–4-aryl ether bonds to produce Hibbert's ketones, demethylation to produce benzodioxanes, and condensation reactions. Unlike lignin model compounds, the condensation of the real lignin in biomass under ALBTH conditions was greatly diminished, possibly due to lignin's remaining in the solid state, limiting its mobility and accessibility of the a-benzyl carbocation to the electron-rich aromatic sites for condensation. This study provided a new approach to effectively isolate depolymerized lignin from lignocellulose in a less condensed form for boosting its downstream valorization.

539017 Catalytic Depolymerization and Liquefaction of Lignin in Ionic Liquid By SO_4^{2-}/ZrO_2 in a Flow through System

Abstract:

Lignin has been considered as a promising sustainable feedstock for high value-added aromatic chemical production. Here, we proposed a novel and e cient process for alkali lignin depolymerization and liquefaction in a flow through system with pretreatment unit, reaction unit and separation unit. In pretreatment unit, lignin was dissolved in N-allyl pyridinium chloride ([Apy]Cl) and SO_4^{2-}/ZrO_2 was used as the catalyst in reaction unit. The effects of different reaction temperatures (150, 170, 190, 210) on the yields of carbon and monomer (on the lignin intake) were investigated, the yields of carbon and monomers appear to be minimized (17.888% and 3.278% respectively) at 150 and increased to 39.179% and 20.254% at 210 respectively. The structural evolution of the alkali lignin was intensively investigated via comparative characterization of raw and residues using FT-IR. The structures of residues were changed with temperature increased, and coke had probably been formed.The thermal stability of [Apy]Cl and regenerated [Apy]Cl were studied. According to the results, during the pretreatment process the β–O–4 linkages were partially cleaved, and in reaction process, SO_4^{2-}/ZrO_2 mainly cleaved the β–O–4, α–O–4 α–O–γ and C–C linkages in lignin and produced phenolic products and their derivatives. Proposed [Apy]Cl pretreatment mechanism of lignin and depolymerization liquefaction mechanism of lignin in [Apy]Cl catalyzed by SO_4^{2-}/ZrO_2 were speculated.

536971 The Impact of Acid Site Concentration and Pore Diameter on the Cracking of Lignin Derived Monomers in Zeolites

Abstract:

In the forest bioproducts industry, tens of millions of metric tons of pulp are produced every year. An important byproduct of the pulping process is lignin, the irregular aromatic biopolymer that acts as a natural glue to hold plants together. Currently, most of the industry burns this complex biopolymer for energy rather than attempt to salvage the valuable aromatic carbon inside. One of the major hurdles to commercializing lignin based organics is that during depolymerization, lignin breaks down into a complex mixture of products. To make lignin valorization an economically feasible process, we must find a way to convert this complex stream into easily separable products.

Using lessons learned from petroleum refining, we show how a mixture of lignin monomers can be converted to an easily separable stream of light hydrocarbons and oxygenated aromatic rings by removing alkyl site chain through cracking reactions. Specifically, this talk focuses on the cracking of propylene from a phenol ring over various types of zeolites. We explore how different aspects of the catalyst, such as the concentration of Brønsted and Lewis acid sites, micropore diameter, and hierarchical structure impact conversion, selectivity, and deactivation, as well as the required regeneration procedure for restoring the original activity. We suggest that strongly bound phenolate species form on the Lewis acid sites, which hinder diffusion of reactants and products into and out of the particles. By selectively removing these sites, we show a change in reactivity and selectivity of the cracking products. Using operando IR spectroscopy, we study how these compounds interact with our catalyst surface and how the various surface species evolve over time, helping us design better and more reliable zeolites.

539104 High Energy Density Fuels Produced from Lignin-Derived Intermediates and Refinery Waste Gas Streams

Abstract:

This paper presents an economic pathway for high energy density fuels production from lignin-derived intermediates and refinery waste gas streams. The lignin-derived intermediates, including toluene and xylenes, are obtained from one-pot lignin conversion catalyzed by metal(s)-zeolite Y catalysts with the presence of hydrogen in water at 250Â°C [1-2]. Alkylation of the intermediates with refinery waste gas streams, which contain up to 10% olefins, and subsequent distillation, produces C8-C12 hydrocarbons with high octane number (>120) and energy density using the existing refinery infrastructure. The octane numbers and energy density of the components are summarized. Brief economic analysis is provided.

[1] H. Wang, H. Ruan, M. Feng, et al, One-Pot Process for Hydrodeoxygenation of Lignin to Alkanes Using Ruâ€Based Bimetallic and Bifunctional Catalysts Supported on Zeolite Y, *ChemSusChem* 2017, 10(8), 1846-1856.

[2] H. Wang, M. Feng, B. Yang, Catalytic hydrodeoxygenation of anisole: an insight into the role of metals in transalkylation reactions in bio-oil upgrading, *Green Chemistry* 2017, 19, 1668-1673.

535560 Towards Valorization of Biorefinery Waste to Polyhydroxyalkanoate: Structural Characterization and Mechanisms

Abstract:

The large-scale production of cellulosic ethanol from lignocellulosic biomass are generating lignins which have high potential to be transformed to sustainable high-value products. Sustainable production of polyhydroxyalkanoate (PHA) using lignin as carbon source provides a promising valorization pathway of lignin. In this study, four lignin streams generated from different fractionation strategies (integrating acid/alkali pretreatment, enzyme hydrolysis, and laccase pretreatment) from corn stover were used in fed-batch fermentation to produce PHA by *P. putida* KT2440. The lignin streams before and after subjecting to the fermentation process as well as the native lignin from corn stover were thoroughly characterized by gel permeation chromatography (GPC), ^{31}P nuclear magnetic resonance (NMR), two-dimensional (2D) heteronuclear single-quantum coherence (HSQC) NMR spectroscopy, and ^{19}F NMR. The GPC data showed that the fractionation processes significantly lower the molecular weight of lignin steams; the polydispersity (PDI) of the lignin streams decreased after the fermentation step with a highest reduction percentage of 50.8%, which may suggest that the lower molecular weight components were more favorable for *P. putida* KT2440 and the molecular weight distribution of lignin streams became uniform. ^{31}P NMR analysis examined the hydroxyl group changes after the fractionation and fermentation processes. Semi-quantitative HSQC data proved that the *p*-hydroxyphenyl lignin may promote the lignin conversion to PHA; more than 78% *p*-coumarate and 15% ferulate structures in lignin steams from three fractionation strategies were consumed, which suggest that consumption of hydroxycinnamates contributed to the production of PHA. ^{19}F NMR analysis achieved the quantification of carbonyl groups in lignin stream; the fermentation generally consumed more than 50.4% aldehyde, ketone, and quinone structures in lignin steam.

538284 Characterization of Deep Eutectic Solvent Extracted Lignin Streams from Endocarp Biomass

Abstract:

Lignin is a promising direct source and contains upgradable building blocks of valuable aromatic chemicals and materials. Endocarp biomass represents a non-edible crop residue in an existing agricultural setting which cannot be used as animal feed nor soil amendment. With considerably higher lignin content and bulk energy density, endocarps have significant advantages to be converted into both biofuel and bioproducts as compared with other biomass resources. As a new pretreatment method, deep eutectic solvent (DES) pretreatment is shown to be highly effective in fractionating lignin from biomass with high yield and purity, in addition to merits of both environmental friendliness and low cost compared with several other pretreatment methods. However, the impact of DES pretreatment on the physical and chemical properties of extracted lignin is so far not clear. Therefore, this study aims to fractionate and characterize lignin streams from endocarps of peach and black walnut using DES in comparison with two conventional pretreatment methods, i.e., dilute acid and alkaline pretreatments. SEM and confocal microscopic analysis revealed distinct structural characteristics of peach and walnut endocarp biomass as compared to switchgrass and woody biomass. Chromatographic and nuclear magnetic resonance (NMR) spectroscopic techniques were employed to understand the lignin composition, structure and depolymerization during DES pretreatment. The resulting lignins were further characterized for purity, molecular weight distribution, and thermogravimetric properties. This study provides a foundation for developing novel conversion techniques for upgrading DES extracted endocarp lignin for value-added chemicals and materials.

538719 Reactivity-Based Fractionation of Lignins Via Reversible Conjunction to Polymeric Amines

Abstract:

Fractionation is a necessary step in the process of upgrading lignins to synthetic materials. Existing methods for separating lignin fractions based on differences in solubility or molecular size generate polymer mixtures with heterogeneous physiochemical properties, which are not suitable for use in high-value plastics or composites. We demonstrate the utility of catalytic oxidation as an approach to partially depolymerizing lignins from diverse sources including raw biomass, enzymatic lignin, and fungal decay lignin. The resulting lignin fragments contain reactive carbonyl groups (i.e. aldehydes and ketones), which conjugate reversibly with primary amines. Using an amine-functionalized polymer as the conjugating reagent, we are able to isolate lignin fractions that are enriched in carbonyl structures. Recovery and reuse of the conjugating reagent are achieved via acid-catalyzed hydrolysis of amine-carbonyl adducts.

538956 Understanding and Modeling Effects of Nitrogen Source on Biosynthesis of Polyhydroxyalkanoates from Benzoate By *Pseudomonas Putida* KT2440

Abstract:

Benzoate has been widely used as a model compound for the study of bacteria catabolism of aromatic compounds. Nitrogen source is critical to the synthesis of intracellular polyhydroxyalkanoates (PHAs) in various bacteria. However, exact effects of nitrogen source in the PHAs biosynthesis process from benzoate in *Psuedomonas putida* is unclear. In this study, the PHAs biosynthesis process from benzoate by *Pseudomonas putida* KT2440 was investigated under nitrogen surplus and limit conditions. In nitrogen limit condition, PHAs accumulated up to 37.3% of its cell dry weight (CDW) and eleven types of monomers were identified through GC/MS analysis. On the contrary, PHAs only accounted for 0.25% of CDW and three main PHAs monomers were found under nitrogen surplus condition. NMR and GPC results confirmed that PHAs molecular weight started to increase when the nitrogen source exhausted while the PHAs concentration maintained at the similar level under surplus nitrogen condition. Kinetic models were developed to adequately predict the biomass growth, PHA accumulation, and benzoate consumption. Results indicated that the cell growth and PHA accumulation fit the Monod-type model and the PHA synthesis process was dependent to nitrogen exhausting condition.

526594 Synthesis and Potential Antiproliferative Activity of Dehydroabietylamine Imidazole Derivatives

Abstract:

ABSTRACT: ABSTRACT: Cancer has attracted distinctive attention from all over the world, and extensive research has been devoted to the development of effective anticancer modalities.[1-2] Imidazole containing compounds have been found to be very important as anticancer, antimicrobial, antibacterial, antifungal, antioxidant, hypotensive, anti-inflammatory and several important therapeutic agents.[3] Recently, a number of bioactive natural products with tricyclic diterpene system have drawn more attention,[4] such as dehydroabietylamine ($\mathbf{L^0}$). A focus of research of dehydroabietylamine derivatives with their anticancer, antibacterial, antifungal, cytotoxic and antiparasitic activities has attracted wide interest in the forest chemical field. In view of the imidazole and dehydroabietylamine derivatives all have considerable anticancer activities, our group hopes to combine them to synthesize and screen target-specific DNA, high anticancer and low toxic dehydroabietylamine derivatives including organic salts ($\mathbf{L^1-L^2}$), amides ($\mathbf{L^3-L^5}$) and Schiff-bases ($\mathbf{L^6-L^7}$). Among these derivatives, several were designed and synthesized as potent anticancer agents. The antiproliferative activities of these compounds were investigated against Hela (cervix), MCF-7 (breast), A549 (lung), HepG2 (liver) human cancer cell lines *in vitro* by MTT assay. It turned out that many compounds showed high activities against these four cancer cell lines and most of these compounds displayed more potent anticancer activities compared with dehydroabietylamine. Besides, our previous work has tested the inducing-apoptosis ability of the compound $\mathbf{L^6}$ and $\mathbf{L^7}$. The convenient strategy may be helpful in the development of future generations of anticancer drugs.

Reference:

[1] C. S. Grasso, Y. M. Wu, D. R. Robinson, X. Cao, S. M. Dhanasekaran, A. P. Khan, M. J. Quist, X. Jing, R. J. Lonigro and J. C. Brenner, The mutational landscape of lethal castration-resistant prostate cancer, *Nature*, 2012, 487(7406), 239–243.

[2] A. Imran, N. L. Mohammad, Y. Haasan and E. Aboul, Imidazoles as potential anticancer agents, *Med. Chem. Commun.*, 2017, 8, 1742–1773.

[3] S. Nishtha, R. A. Mohammad and B. Sandhya, Imidazole: An Essential Edifice for the Identification of New Lead Compounds and Drug Development, *Mini-Reviews in Medicinal Chemistry,* 2018, 18, 142-163.

[4] M. D. A. Auxiliadora, B. R. Pablo, B. F. Francisco, A. G. Miguel, Synthesis and antileishmanial activity of C7- and C12-functionalized dehydroabietylamine derivatives, *Eur. J. Med. Chem.*, 2016, 121, 445-450.

529843 Protein Content and Amino Acids Profile in Ten Cultivars of Ginkgo (*Ginkgo bilona* L.) Nut from China

Abstract:

As a traditional food and medicine source, ginkgo (*Ginkgo biloba* L.) nut is popularly consumed in East Asia. The aim of this work is to characterize protein content and amino acids profile in ten ginkgo nut cultivars, named successively as No. 1 to No. 10. There were observed differences among the cultivars with respect to the contents of protein and amino acids, except Cys. The No. 6 cultivar presented the highest protein content (22.13 g/100 g DW), while the No. 9 did the lowest (16.17 g/100 g DW). The contents of EAA and NEAA were revealed to vary in the range of 14.32 to 26.22 and 21.38 to 41.07 g/100 g protein, respectively. The most abundant EAA was Leu, and the first limiting amino acid was Lys. The level of Arg was attractive, especially in the No. 5 cultivar where it (1741 mg/100 g DW) is comparable to hazelnut and pistachio. As confirmed by AAS and EAAI, the No. 5 cultivar presented the best amino acids profile and protein quality amongst these cultivars. These results have relevance to the possible utilization of ginkgo nut in food industry and direct consumption in a wider range.

534341 Hydrothermal Treatment of Paper Mill Sludge: Nutrient Characterization

Abstract:

Each year the pulp and paper industry generates large amounts of effluent treatment sludge. The sludge is made up of three different fractions namely, primary, secondary, and tertiary sludge. Depending on the layout of the sludge handling in the individual mill, these three sludges can either be extracted separately or mixed into a stream called mixed sludge. Usually secondary and tertiary sludge are mixed together. Paper mill sludges contain several plant nutrients including P, K, N, Ca, and Mg. However, the nutrient, heavy metal, and organic composition of sludges is determined by the fiber source and treatment process, and is quite varied.

Hydrothermal carbonization (HTC) has received considerable attention during the last several years as a prominent alternative for extracting nutrient from different biomass feedstock like forest residues and industrial or municipal wastes. One of the main advantages of HTC is the use of aqueous solutions under high temperature (180-260 °C) and autogenously generated pressure (5-50 bar). As a result, moist and wet feedstocks are favorable for HTC.

This goal of this study was to determine the effects of HTC conditions on the nutrient contents in paper mill sludge. A 600 ml Parr batch reactor was used for HTC. Reaction temperature was varied from 180-260 °C, while reaction time was varied from 5-30 min. The nutrient content in solid hydrochars were assessed by evaluating ultimate analysis, FTIR, and NMR. Meanwhile, the process liquid is characterized with GCMS, HPLC, UV-VIS, NMR, and pH. The results suggested that the nutrient content of hydrochar and process liquid increased with reaction temperature. However, residence time has insignificant effect on it.

536686 Inhibitory Effect of Biomass Hydrolysates on Glucose Transport in Microbial Fermentation

Abstract:

A wide range of toxic compounds was generated from lignin and sugars degradation in biomass pretreatment. These hydrolysate inhibitors dramatically inhibit cell growth and production rate in microbial fermentation. Phenolic aldehydes and ketones have been suggested to the most toxic compounds to microbes. However, the mechanism of inhibition has not been elucidated. Glucose transport across the yeast cell membrane typically is mediated by specific membrane carriers. The effect of hemicellulose hydrolysates on glucose transport and cell-free fermentation with yeast was studied. Unexpectedly, it was observed that cell-free yeast extracts consumed all the glucose in the undetoxifed hydrolysates within 12 h and produced 9.5 g/l of ethanol, but the intact yeast could not ferment the undetoxifed hydrolysates. It was also noticed that no intracellular glucose was detected in yeast after incubation with the undetoxifed hydrolysates, while 46 mg/g intracellular glucose was presented inside of yeast after incubation with the detoxified hydrolysates. The volumetric ethanol productivity was correlated very well with the intracellular glucose concentration. This study indicated that the toxic compounds mainly inhibit the glucose transport step, but not the fermentation step in yeast cells.

537219 *Effects of P*-**Hydroxybenzoic Acid and 2-Napthol** *on Dilute Acid Pretreatment of Aspen*

Abstract:

Biomass pretreatment is essential for producing biofuels from lignocellulosic biomass. The aromatic additives (such as *p*-hydroxybenzoic acid and 2-napthol) have the potential to prevent lignin repolymerization in pretreatment and improve enzymatic hydrolysis. Effects of *p*-hydroxybenzoic acid and 2-naphthol additives on dilute acid pretreatment were investigated. In this study, we examined the enzymatic hydrolysis of the pretreated substrates and the physicochemical properties and structures of residual lignin. It was observed that *p*-hydroxybenzoic acid and 2-naphthol improved the 72h hydrolysis yield of dilute acid pretreated substrates by 13% and 17%, respectively. Langmuir adsorption isotherms showed that *p*-hydroxybenzoic acid and 2-naphthol decreased the residual lignin affinity to cellulase enzymes and reduced the distribution coefficient from 0.232 L/g to 0.124 and 0.084 L/g, respectively. The residual lignin molecular weight was determined by gel permeation chromatography. The potential reaction between lignin and *p*-hydroxybenzoic acid and 2-naphthol was characterized by HSQC-NMR. The study indicated that aromatic additives could change lignin hydrophobicity and reduce lignin condensation in biomass pretreatment and enhance the enzymatic hydrolysis.

537638 Investigating the Sorption Capacity of Hydrochar for Organic Pollutants and Comparing with That of Powdered Activated Carbon (PAC) As a Method of Treating Contaminated Water

Abstract:

Water is perhaps the most critical component of our diets. Water is plentiful on earth, but obtaining clean potable water for human consumption is a big challenge in many parts of the world. Water is contaminated by organic compounds coming from industrial waste (e.g. pharmaceutical chemicals) and from agricultural water run offs (e.g. pesticides, herbicides, chemical fertilizers). These chemicals have deadly effects on human bodies if consumed in the amount exceeding the safe-level limits. One of the most common methods of removing organic pollutants is adsorption, a phenomenon in which absorbable solute (adsorbate) is chemically or physically attached to a solid with highly porous surface structure (adsorbent).

Due to its high level of microporosity which leads to high surface area (> 800 m^2/g), powdered activated carbon (PAC) is the most common adsorbent being employed to remove organic chemicals from water. PAC is usually produced from carbonaceous (biomass) sources via dry pyrolysis. The shortcoming of this process is the requirement for the biomass feedstock to have extremely low moisture content. Drying biomass materials requires a large amount of energy, especially if the materials contain high moisture content.

This study aims to evaluate the potential of hydrochar as a low-cost sorption agent. Hydrochar is solid carbon material produced from a thermochemical process, namely hydrothermal carbonization (HTC), in which wet organic materials (e.g. food wastes) are immersed in water and heated to temperature between 180 and 270^{0}C under sub-critical condition during which water remains in liquid phase. The absence of liquid-to-vapor phase change of water makes the HTC process less energy intensive compared to the process of producing PAC which involves water vaporization and high-temperature carbon activation.

It has been reported that, similar to commercial PAC, hydrochar produced from HTC process also displays high level of microporosity. This property brings up the question whether, like commercial PAC, hydrochar will have the high sorption capacity and potential to become a low-cost sorption agent. Since it can be created by common organic materials like food wastes, hydrochar can be produced anywhere using local organic resources, even in rural and developing communities. Another advantage of hydrochar production is the reduction of organic wastes flowing to landfills, thus reducing the potential synthesis of greenhouse gases[4].

In this sorption study, a large variety of hydrochar materials are evaluated for their sorption capacities. The hydrochars were made from different types of food waste (carbohydrates, vegetables, plant proteins and animal proteins) at various HTC process temperatures and retention times. The hydrochars are tested for adsorbing three different common organic pollutants, namely 2,4 Dichlorophenol (2,4 DCP), acetaminophen, and Bentazon 4 herbicide. To be reported are the effects of types of food feedstock and the HTC process conditions (temperatures and reaction time) on the sorption capacities of hydrochars for the three organic pollutants tested along with the comparison of hydrochar sorption capacities with respect to that of the commercial PAC.

*Contact: Dr. Justinus Satrio (justinus.satrio@villanova.edu)

538004 Synthesis of Hardwood Lignin Model Polymer and Its Effect on Enzymatic Hydrolysis of Cellulose

Abstract:

Lignin is an abundant and ubiquitous component in lignocellulosic biomass. Dimeric lignin model compounds have been often used to investigate lignin degradation and catalytic conversion. However, these models are difficult to represent the three dimensional complex structure of lignin polymer and their effects on enzymatic hydrolysis of cellulose.

The lignin model polymers were synthesized to contain β-O-4 and β–β linkages that are structural motifs ubiquitous in native lignin obtained from hardwood. The method mainly consists of 4 parts: 1) preparation of monomeric acetate, 2) multiple-step synthesis of β–β dimeric diacetate, 3) construction of β-hydroxyl ester-containing oligomers via lithium diisopropylamide (LDA)-promoted aldol type reaction of β–β dimeric diacetate with monomeric acetate, and 4) reduction of the ester moiety with $LiAlH_4$ to give the lignin model polymer. The combined yield of the last two steps is 27%. The β–β linkage is from β–β dimeric diacetate while the β-O-4 linkage is the repeating unit of this polymer.

The synthesized lignin polymers were characterized by GPC analysis and HSQC NMR. The number average degree of polymerization (DP_n) of the polymer is 8.25 and the molecular weight of lignin polymers were 2378. These artificial lignin polymers decreased the 72 h hydrolysis yield of Avicel from 57% to 53%, which is similar to the effect of native lignin on Avicel.

539085 Extraction and Recovery of Sinapic Acid from Oleaginous Biomass (Mustard Bran): A Sustainable Access to a Valuable Phenolic Platform Chemical

Abstract:

The objective of this research project was to develop a cost-effective and sustainable integrated process for the extraction and recovery of sinapic acid from oleaginous â€œoil-bearingâ€ biomass as a value added chemical alongside cellulose and lignin. Waste products like mustard bran and canola meal are rich in sinapine (~15 mg/g), a derivative of sinapic acid, which has the potential to serve as a value-added product for applications in the cosmetic, the plastic and the pharmaceutical industries due to its antioxidant and anti-UV properties.

By employing an optimized chemo-enzymatic technique, sinapine was effectively liberated and completely converted into sinapic acid. Following this, the sinapic acid was recovered using an extensively optimized membrane filtration technology involving nanofiltration and diafiltration. Lastly, the residual biomass (fats, cellulose, lignin) were also fractionated and characterized for their use in other products. With these unique approaches in fractionation, separation and process integration, the process developed has an improved efficiency, cost effectiveness and environmental impact. The establishment of these processes can lead to new technology developments, and economic opportunities, which can, in turn, enable the cost-effective production of advanced bioproducts.

539741 Process Design for Conversion of Coconut Coir Pith to Bioplastic and Byproducts

Abstract:

Coconut Coir Pith (CCP) is a widely available biorenewable resource especially in coconut growing regions of the world. Coconut coir is the fibrous material that makes up a coconut husk and can be used to make variety of products.This is a rich source of lignocellulose which can be converted to value added renewable bioproducts. Bio Plastics such as Poly Hydroxy Alkanoates (PHA) are sustainable plastic materials that are bio degradable and can be manufactured from biorenewable resources. It is possible to use coconut coir pith as a raw material for PHA thus providing significant value addition to coconut coir industries. Here we present a process design and technical and economic analysis of effectively converting the waste CCP into useful bioplastics and lignin biopolymer residue. Bioplastic (PHA) can be obtained from the cellulose and hemicellulose in CCP and the residual lignin biopolymer can be used for other applications. The process steps for the conversion of CCP include: Pretreatment of CCP to improve accessibility of cellulose; Enzymatic hydrolysis of pretreated CCP to produce fermentable sugars; Bioconversion of the sugar stream into bioplastics using genetically engineered proprietary strains to give PHAs at high yields and tailored quality; Separation and purification of PHA bioplastic as well as solid biomass residue containing lignin biopolymer as a value added byproduct. Process design and analysis include: Process flow diagram and material and energy balances of the overall process system including various unit operations; Process equipment and sizing; Economic analysis showing profitability and recommendations for small scale facility.

532652 Selecting Solvents for Lignin Value Prior to Pulping

Abstract:

Solubilization of lignin and its subsequent valorization are the new frontier for chemicals and products from renewables. Lignin valorization enables favorable economics for a biorefinery, but the efficacy of an organic solvent must be paired with a set of economic and process-based considerations. We describe a methodology for selecting organic solvents for biomass treatment with a specific focus on lignin value prior to pulping (LVPP), a next-generation pulping system. For treatments that do not fiberize or solubilize biomass, we describe elimination and selection criteria that provide a directed search and validation of viable organic solvents. We highlight the economic drivers behind the environmental, process, thermodynamic, and efficacy criteria.

540754 A Study on Extent of Chain Crosslink on HDT Improvement ofÂ Poly (lactic acid)

Abstract:

The low heat deflection temperature (HDT) of poly(lactic acid) (PLA), at approximately 55°C, has greatly limited its use in many fields of work especially in the high temperature applications. It was anticipated that the molecular structure of PLA greatly influenced the HDT values. Therefore, samples of crosslinked PLA with different crosslinking density were prepared by reactive extrusion in the presence of triallyl isocyanurate (TAIC) and luperox 101 to observed the impact of changes to the molecular structure on the HDT of PLA. The gel contents, rheological properties, crystallinity, HDT and dynamic mechanical (DMA) properties of the crosslinked PLA were determined in this study. The results found that the HDT of the crosslinked PLA remained unchanged around 55 °C, even when the sample was highly crosslinked. The DMA studies suggested that the glass transition of the crosslinked chains were the same as the linear PLA. It was then inferred that the chain segment movement had not been limited by the crosslinking, and thus the final HDT of the amorphous PLA was not improved. Finally, we increase the crystallinity of the crosslinked samples, finding that the HDT was improved at a critical crystallinity value. The research concluded that the improvement of the HDT of PLA can only be achieved by limited the movement of the PLA chain segments, rather than the chain coil, which can be a guide in our future work on improving the HDT of PLA.

Acknowledgment: This research is financially supported by the Ontario Research Fund, Research Excellence Program; Round-7 (ORF-RE07) from the Ontario Ministry of Research, Innovation and Science (MRIS) (Project # 052644 and 052665); the Ontario Ministry of Agriculture, Food and Rural Affairs (OMAFRA)/University of Guelph Gryphonâ €™s LAAIR Program (Project # 298636); Agriculture and Agri-Food Canada (AAFC) and Competitive Green Technologies through AgriInnovation Program project (Project # 052882 and 051910); and the Natural Sciences and Engineering Research Council (NSERC), Canada Discovery Grants (Project # 401111 and 400320).

531930 Effect of Lignin on Nanofibrillated Cellulose Production

Abstract:

Nanofibrillated celluloses (NFCs) have shown many potential applications due to their biodegradability, high surface-to-volume ratios and high mechanical properties. The NFCs can be produced through the mechanical defibrillation of cellulosic fibers with or without a chemical or enzymatic pretreatment. However, the NFCs trend to form cellulose bundles through the inter-fibrillar hydrogen bonding. Moreover, the mechanical defibrillation can cause the thermolytic scissions of the glycosidic bonds of cellulose and concomitantly create cellulose radicals. The formed cellulose radicals can react each other and result in the undesirable recombination of the NFCs. As a result, the NFCs yields are typically low.

Lignin is a free radical scavenger, and therefore can stabilize the cellulose radicals through forming the very stable phenoxy radicals. On the other hand, lignin is hydrophobic, and thus can interrupt and reduce the inter-fibrillar hydrogen bonding once it attached to the NFCs surface. So, lignin could potentially counteract the recombination reactions between the highly reactive cellulose radicals and prevent the formation of cellulose bundles. Herein, this study investigated the effects of lignin on the NFCs production. The results clearly showed that lignin greatly facilitated the mechanical defibrillation of cellulosic fibers and thus improved the NFCs yields. And lignin also reduced the diameters of the NFCs.

531760 Cellulose Nanocrystal: Synthesis, Characterization, Dispersion in Organic Media and Surface Modification

Abstract:

Abstract: The Cellulose Nanocrystal (CNC) surface is highly hydrophilic, which hinders its dispersion in organic polymers. However, successful modification of CNCs often requires reactions in apolar organic medium. Here, different polar aprotic solvents, such as acetonitrile, dimethyl sulfoxide, propylene carbonate and acetone, were examined to disperse CNC. Aqueous CNC suspensions at different pHs were freeze dried and redispersed in the organic solvents. From the visual inspection of vials and UV-vis spectra of each CNC suspension of different solvents it was found that dimethyl sulfoxide was the best solvent among them to disperse the CNC and make a stable suspension. Moreover, dimethyl sulfoxide was later used as a reaction medium for the oxa-Michael addition of the primary hydroxyl groups on the CNC surface to N-Benzylmaleimide. Herein, two reaction routes were explored. First, an acid catalyzed method was used where the sulfate ester groups in freeze dried CNC themselves were used as acid catalyst at 100°C. In the second case, the reaction was run in basic medium catalyzed by triethylamine at 70°C. The reaction samples were examined in NMR and FTIR before and after the reaction indicating successful modification in both cases; however the TEA catalyzed reaction resulted in a higher degree of modification. No efforts were made to further remove water other than freeze drying the CNC and redispersing in DMSO.

CNCs for this work were isolated in-house from southern bleached softwood kraft (SBSK) pulp from Mississippi woods by sulfuric acid hydrolysis, as it provides highly stable negatively charged and acidic nanocrystal aqueous dispersions. The size of CNC was measured using AFM and light scattering resulting in an average length of 122 ± 70 nm and height of 6.1 ± 3 nm. Quantification of superficial sulfate groups was performed by conductometric titration with sodium hydroxide; the sulfur content in CNC was found to be 0.41 wt%. The CNC concentration was measured by total organic carbon (TOC), UV-vis spectroscopy and gravimetric techniques.

Results from this work are expected to allow further chemical modifications of CNC with the goal of introducing them in thermoset nanocomposite resins. To the knowledge of the authors, this report is the first time an oxa-Michael addition has been used to modify the surface of CNC.

532937 Liquid-Liquid Lignin–Solvent Systems: Phase Behavior, Characterization and Applications

Abstract:

Lignin is unique among biopolymers in having significant aromatic character, which makes it potentially useful for a wide range of applications, including electrodeposited coatings and carbon fibers. Unfortunately, most of the commercial-grade lignins available today (primarily Kraft lignins) are too polydisperse and have a high metals and ash content, eliminating them from contention for most high-value applications.

Thies and co-workers [1] have developed a process for simultaneously solvating, fractionating, and purifying the lignin polymer recovered from the biomass by-product streams of pulp-and-paper mills and lignocellulosic biorefineries. Called Aqueous Lignin Purification with Hot Acids, or ALPHA for short, the process involves combining solid lignin with organic solvent–water mixtures to produce two liquid phases: a highly solvated, purified polymer (lignin)-rich phase containing higher molecular weight(MW) lignins and a salty solvent-rich phase containing lower MW lignins. Recently, our group [2] found that carbon fibers spun from purified, fractionated high molecular weight lignin has more than doubled tensile strength and higher tensile modulus than any lignin-derived carbon fiber reported in literature. However, this complex lignin-solvent system is lack of fundamental studies to help understand why ALPHA is so effective for processing lignin.

Liquid–liquid equilibrium (LLE) phase behavior has been measured for two different renewable solvent systems, with the results being presented as ternary phase diagrams depicting all regions of both fluid and solid phase behavior. A number of important thermodynamic properties relevant to the processing of lignin can be obtained from these diagrams, including the size of the LLE region available for polymer processing, and the distribution of lignin between the polymer-rich and solvent-rich phases. The molecular architecture of lignin molecules from each phase were measured both by Gel Permeation Chromatography (GPC) and Static Light Scattering (SLS). Finally, Heteronuclear Single Quantum coherence(HSQC)-NMR was used to measure the functionality changes of ALPHA-processed lignin fractions for investigating proposed reaction mechanisms.

[1] Klett, A. S.; Chappell, P. V.; Thies, M. C., Recovering ultraclean lignins of controlled molecular weight from Kraft black-liquor lignins. *Chemical Communications* **2015,** *51* (64), 12855-12858.

[2] Jin, J.; Klett, A. S.; Ding. J.; Thies, M. C.; Ogale, A. A., Carbon Fibers Derived From Fractionated-Solvated Lignin Precursor with Enhanced Mechanical Performance. in preparation, 2018.

535581 Preparation of pH-Responsive Latex Films from Glycerol Based Dendritic Precursors for Food Packaging

Abstract:

Abstract: Food safety is an important issue and a high priority globally placing human and animal health. Food spoilage causes approximately 50% waste of our food before and after it reaches to the customers due to the delay in food spoilage detection. It has been observed that the spoilage of food occurs in stages one indicator of which is a variation in pH as the food material undergoes changes in palatability. This study aims to encapsulate pH-sensitive and food-safety dye (e.g. alizarin S dye) in barrier biopolymers/films with natural and/or plant-based origins via miniemulsion polymerization. The glycerol-based acrylate dendrimer monomer is proposed to substitute synthetic butyl acrylate in the barrier film recipes to provide both pH-sensitivity and film physical properties. The achievement of nanoencapsulation provides the function of precise pH-sensitivity for barrier food packaging applications. At a varying pH, the nanoshell thickness and structure can be tuned to slowly release of food-safety dye for food spoilage detection. Transmission Electron Microscopy (TEM) images in combination with zetasizer analysis data indicate uniform size distribution of polymer particles, successful dye encapsulation in stable latex and dye release upon pH variation validated by fluorescence spectrometry. The properties of bionanocomposite films will be characterized via a variety of instruments such as Differential Scanning Calorimetry (DSC), tensile machine, and Gel Permeation Chromatography (GPC). We further evaluate the film visual color contrast due to different pH and its application on the pork sample.

539259 Hydrothermal Carbonization of Biomass: Examination of Post Synthesis Treatment and Characterization Techniques

Abstract:

Hydrothermal carbonization is a thermal treatment process during which biomass is converted to a carbon-rich, highly aromatic and chemically active solid (hydrochar). The solid produced is useful as a solid fuel, fertilizer, soil amendment, gas storage, and water purification. In this work hydrothermal chars formed from simple sugars have been modified via the addition of acid during synthesis, the addition of acid after synthesis, the addition of nitrogen containing compounds and via ball-milling. The structure of hydrochar was then investigated using Infrared spectroscopy, NMR spectroscopy, elemental analysis and Raman spectroscopy. To date, Raman spectroscopy has been an underutilized technique for hydrochar characterization, despite the sensitivity of the method toward the carbon-carbon bonds that compose hydrochar. The main reason for the underuse of Raman spectroscopy is the lack of a model to interpret hydrocharâ€™s complex, overlapping spectrum. We have determined that the characteristic D and G band of carbonaceous materials is an incorrect interpretation for hydrothermal chars, and is resultant from thermal changes in the material and not from how the material is synthesized. In this work a new method for taking the Raman spectra of hydrothermally treated carbons is demonstrated. We then apply fitting methods derived Density Functional Theory to examine the structure of the molecules, taking a brief look at how the different treatment methods listed above change the furan to arene ratio of the material.

539366 Development of Bioplasticizers from Soybean Oil

Abstract:

Plasticizers are high-boiling organic liquids that impart flexibility to polymers. Polyvinyl chloride (PVC), the third largest commercial plastic, is the primary polymer that utilizes plasticizers. The most commonly used plasticizers are a class known as phthalate esters. Recently the use of phthalates in products such as medical plastics, food contact materials, and childrenâ€™s toys has come under intense scrutiny as they have been shown to cause negative health affects including hormonal disruptions, carcinogenic properties and developmental effects. Vegetable oils, especially soybean and high oleic soybean, are particularly suitable raw materials for developing new plasticizers due to their ready availability, low cost and ease of chemical modification. Numerous functional groups were introduced at the ester head group and on the back bone of the fatty acid esters to create changes in polarity and branching to enhanced PVC compatibility. The soybean oil triacylglycerols were converted to fatty acid esters by transesterification with various alcohols such as pentanol and glycerol, while the double bonds on the back bone were converted to epoxy function. The epoxy groups were ring opened with acetic acid to form estolide functions on the fatty acid back bone. The free hydroxyls were acetylated to give the final products. The resulting fatty acid esters showed excellent physical and plasticizer properties. They have low color, viscosities, and acid values. The plasticizers compounded with PVC gave low gelation times and temperatures with high thermal stability. The best performing bioplasticizers synthesized would be useful as replacements for phthalate plasticizers as they are non-toxic and perform equal to or better than current phthalate plasticizers.

542620 The Impact of Shale Gas and Oil on the Chemical Industry

Abstract:

Lower alkanes and aromatics are very useful as both fuels and chemical raw materials. Occasionally supply disruptions constrained availability but also encouraged investigation of alternative feedstocks and chemistries. Recent new technologies for the production of unconventional gas and oil sources from tight shale formations significantly increased availability of both methane and condensate, radically altered process economics, resulted in the construction of much new intermediates capacity, and multiplied the ways traditional and alternative chemistries might be exploited. This presentation will outline historically how natural gas components became useful as chemical feedstocks, how government regulations occasionally led to perturbations in availability and price, how supply crises led to the development of alternative chemistries, catalysts, and ultimately the exploitation of shale resources. More specifically, will be discussed how excess coproducts, especially condensate, from increasing shale gas and oil production for energy are having significant unexpected impacts on the chemical process industries and opportunities for new process chemistry research.

541186 Sustainable Energy and Chemicals: Past, Present, and Future

Abstract:

The last 40 years has seen the U.S. energy situation vacillate from gasoline rationing and concerns over Peak Oil, to projected energy and feedstock prices that were noncompetitive in global markets for chemicals production, to the economic revolution of frack gas and light tight oil. The importance of low cost energy and feedstocks in driving economic opportunity will be discussed relative to this history, along with the importance of sustainability in decision making to gain stakeholder buy-in and acceptance. Strategies for managing career paths and competencies despite changing targets and high uncertainty will be addressed.

542618 Disruptions: What the Future May Hold

Abstract:

Disruptive forces are all around us. They change the way we communicate and work. They change the way we develop goods and services. They also change the types of products that are manufactured. Examples such as the way in which we produce and consume energy or the feedstocks we use to produce product have seen many changes over the course of the last few decades, and these disruptions are expected to continue. The chemical engineering discipline and the expectations for a chemical engineerâ€™s skills has also changed and adapted as a result of these disruptive forces. A discussion of the broad themes associated with these forces and the implications that they may have to chemical engineers in the future will be discussed.

542621 Geopolitical Factors Influencing the Evolution of the Chemical Industry

Abstract:

The basic chemicals and plastics industry is built on a simple principle: find a cheap feedstock and convert it into something more valuable. However, the right feedstock depends on place and time. Certain policies and actions of various governments have influenced the evolution and profitability of the chemical industry during the past 50 years, sometimes in surprising ways. For example, high oil prices and US government funding encouraged the development and commercialization of coal gasification technology during the late 1970s and early 1980s. This enabled China to develop a competitive basic chemical industry using coal as a feedstock during the last 25 years. Similarly, an abrupt increase in oil prices during the first decade of the 21st century encouraged the present shale gas revolution, causing a major shift in the economy of producing basic chemicals and polymers in the US. If history is our guide, then what governments choose to fund and regulate may be as important as market forces in determining the shape of our industry and profession in the future.

542622 Agility & Resilience: How to Maintain Career Competitiveness in the Changing Chemical Industry

Abstract:

The world around is changing. The shift to protectionism, global instability and price volatility of raw materials and feedstocks has the potential to wreak havoc on the chemical industry. With the revolution of shale gas in North America and the sudden drop in feedstock costs, the chemical sector in the US has reasserted its historic dominance. Learn how the role of a chemical engineer can contribute to industry growth and resilience, while adapting to allow for personal agility throughout the career life cycle.

530038 Experimental and Computational Studies of Interfacial Interactions of Lignin Dimers with Lipid Bilayers

Abstract:

Experimental and Computational Studies of Interfacial Interactions of Lignin Dimers with Lipid Bilayers

Mahsa Moradipour[1], Xinjie Tong[2], Poorya Kamali[3], Shardrack O. Asare[3], Bert C. Lynn[3], Dorel Moldovan[2], Stephen E. Rankin[1], Barbara L. Knutson[1]

[1]Department of Chemical Engineering and Materials Science, University of Kentucky

[2]Department of Mechanical and Industrial Engineering, Louisiana State University

[3]Department of Chemistry, University of Kentucky

Lignin is an organic substance present in cells, cell walls, and between the cells of all vascular plants. While comprised of only three monolignols (aromatic alcohols), it has a complex structure variation in both sequence and types of bonds between monomers. Due to the heterogeneity of the lignin structure, the goal of deconstructing lignin to commercially viable small molecules for use as chemicals or advanced materials has been elusive. One potential application of lignin derived molecules is antimicrobials or antifouling coatings. This application is based on the known inhibitory nature of deconstructed lignin components to microorganisms.

In this work, the interaction of three Beta-O-4 dimers of coniferyl alcohol (GG lignin dimer) with synthetic lipid bilayers was studied: a commercially available compound with no alcohol tail, the ÒnaturalÓ dimer containing a hydroxyl tail, and a benzyl-modified dimer. Small lignin molecules are thought to act as antimicrobials by interacting with cell membranes, represented here as model phospholipid bilayers. A quartz crystal microbalance (QCM) was used to study the amount of dimer uptake on supported lipid bilayers and to compare effect of dimers on the properties of the bilayer. Corresponding differential scanning calorimetry measurements (DSC) using lipid vesicles were used to more directly study the change in bilayer phase behavior due to incorporation of dimers into the bilayer. These measurements suggest that the commercial GG and the benzyl-modified GG enter the hydrophobic region of the bilayer, with similar partition coefficients (log $K_{m/w}$) of 2.9 and 3.0, respectively. However, significantly less natural dimer entered the lipid bilayer (log $K_{m/w}$ = 2.10).

Using molecular dynamics simulations (MD) and free energy calculations, the interaction, absorption, and partitioning behavior of lignin were investigated. The MD simulations show that aqueous monomers and dimers absorb rapidly to the bilayer-water interface. The profile of the free energy (PMF) curves indicate that the absorbed monolignols are located close to the water lipid bilayer interface with very little affinity for the bilayer interior. In contrast, some dimers are able to penetrate significantly into the interior of the bilayer. In addition, MD simulations of bilayers in the presence of higher concentration of dimers (10%, 20%, and 40%) shows that they can have strong effect of bilayer structure and permeation properties. The commercial and benzyl-modified dimers are both more hydrophobic than the natural dimer and therefore have higher affinity for bilayer interior and have a greater effect on its structure. Overall, our results indicate that minor differences in structure in lignin-based derivatives have a significant impact on their interaction with lipid bilayers and that the techniques used in this study will provide fundamental insight into interactions for different lignin oligomers.

á This project is supported by NSF EPSCoR Track-2 RII, Award No. OIA1632854, and National Science Foundation REU Program, University of Kentucky (grant no. 1460486

548514 Understanding Barrier Properties of Plastics for Packaging Applications: A Comprehensive Approach to Polymer Performance

Abstract:

Moisture, carbon dioxide, and oxygen in air produce spoilage of food at a fast rate. Packaging, based on plastics, is designed to increase the shelf life by creating a barrier between food and air. Packaging also protects food from other contaminants including bacteria. Plastic packaging has become widespread and has seen overtime high success due to the low cost to transport products compared to those packaged with metals or ceramics. Plastics are light weight and have high performance for multiple applications including protecting food and innumerable merchandise. Plastics, including petro-based or bio-based, are all different and many of them make excellent light weight, transparent, and though packaging, while others are opaque and strong and/or rigid. Different plastics present different barrier properties and are selective towards moisture, oxygen, or carbon dioxide. Packaging is normally layered due to the different properties of plastics. Measuring and understanding the properties of plastics is essential. Molecular weight, molecular orientation, chemistry, and crystallinity are all essential to understand the overall phenomenon. We will discuss a correlation between these variables with the barrier performance of various plastics including bio-based polymers. It has been found that bio-based polymers can perform as well as petro-based polymers based on barrier properties. Barrier properties of polymers is a complex phenomenon that depends not only on the crystallinity, but on the chemical structure and properties of the polymers. Layering of polymers is an excellent technical recourse to improve the barrier properties of plastics.

Acknowledgements:

This research is financially supported by the Ontario Ministry of Agriculture, Food and Rural Affairs (OMAFRA), Canada/University of Guelph - Bioeconomy for Industrial Uses Research Program Theme (Project # 030255); the Natural Sciences and Engineering Research Council (NSERC), Canadaâ˘'Discovery Grants (Project # 401111 and 400320); and the Ontario Research Fund, Research Excellence Program; Round-7 (ORF-RE07) from the Ontario Ministry of Research, Innovation and Science (MRIS) (Project # 052665 and #052644).

548515 Itaconated Epoxidized Corn Oil: a Fully Biobased and Non-Volatile Alternative of Acrylated Epoxidized Soybean Oil

Abstract:

In this paper, monomethyl itaconated epoxidized corn oil (MIECO) was synthesized by melt ring-opening esterification of monomethyl itaconate (MI) with epoxidized corn oil (ECO). Downstream corn oil, which obtains as a by-product during production of ethanol, was used to synthesize ECO by reported epoxidation method. The chemical structures of MI, ECO, and MIECO were characterized and confirmed by differential scanning calorimetry (DSC), Fourier-transform infrared spectroscopy (FTIR), and ^{1}H Nuclear Magnetic Resonance spectroscopy. The isothermal thermogravimetric analysis (TGA) at different temperatures (30, 60, and 90 °C) indicated that MI demonstrated tremendously low volatility even at high temperature (90 °C), while acrylic acid showed very fast volatility even at low temperature (30 °C), which indicated that MI is excellent green alternative to acrylic acid. The thermal and mechanical properties of as-synthesized MIECO were compared with commercial acrylated epoxidized soybean oil. The cured pristine MIESO showed similar properties to the pristine acrylated epoxidized soybean oil. Therefore, as-synthesized MIECO provides a fully biobased alternative to commercially available acrylated epoxidized soybean oil, which suffers from invasive odor, corrosive and toxic in nature.

Acknowledgements: The authors would like to thank the following for their financial support: Ontario Ministry of Agriculture, Food and Rural Affairs (OMAFRA)/University of Guelph - Bioeconomy for Industrial Uses Research Program (Project # 030251); and the Ontario Research Fund, Research Excellence Program; Round-7 (ORF-RE07) from the Ontario Ministry of Research, Innovation and Science (MRIS) (Project # 052644 and 052665).

540656 3D Printing of Modified PA11 and Biocomposites: Processing and Performance Evaluation

Abstract:

3D printing is an innovative technology which has matured substantially over the last decade, evolving from a hobbyist niche into a real industrial tool. The greatest advantages of 3D printing with the fused deposition modeling (FDM) method include its low start-up cost, minimal footprint, rapid turnaround time, customizability and design control. These aspects make this technology extremely well suited for biomedical applications, where there is no â €œone size fits allâ€ solution, as patients require highly individualized treatment and care. Using 3D printing, biomedical devices can be produced on a custom basis for individual patients. This research investigates the FDM processing of polyamide 11 (PA11), a flexible and durable polymer which can be made with 100% bio-based resources. PA11 has been used successfully in selective laser sintering (SLS), but has not been applied in FDM at this point. In this project, models were successfully prepared using FDM printing of a biocompatible (UPS Class VI certified) PA11. Printable filaments were prepared using a Leistritz twin screw extruder. In order to improve the suitability for printing and processability into filaments, different chain extension and reactive blending strategies were employed. In order to improve thermomechanical behaviour, the production of PA11 biocomposite filaments were also investigated.

Acknowledgements:

This research is financially supported by the Ontario Ministry of Agriculture, Food and Rural Affairs (OMAFRA), University of Guelph, Bioeconomy Industrial Uses Research Program Theme Project #030252; and the Natural Sciences and Engineering Research Council (NSERC), Canada Discovery Grants Project #400320.

533418 Biodiesel As a Green Solvent to Improve the Dilute Acid Pretreatment of Lignocellulosic Biomass

Abstract:

Cellulose hydrolysis has been found to be significantly retarded, particularly at high severity acid treatment conditions, principally due to the formation of pseudo-lignin. In order to suppress its formation during hydrothermal and/or acid pretreatment, a strategy assisted with a green solvent i-e biodiesel employing with different acids has been evaluated in this work. The qualitative analyses revealed that indeed biodiesel was effective enough to considerably suppress the pseudo lignin formation by way of dissolving out the furfural formed during high severity acid pretreatment conditions. This was attributed to the characteristics of biodiesel which it has to solubilize the expected fractions of furfural formed during typical acid pretreatment and contributing thus to minimize the pseudo lignin formation. Following pretreatment, the samples were subjected to SEM images to study the surface morphology in order to visually verify the evidence of suppression in pseudo lignin formation. Furthermore, XPS, XRD and FTIR analyses were also carried out to confirm the structural changes during this pretreatment. All of these conclusively suggest that undoubtedly biodiesel is an effective green approach to help improve the acid pretreatment of lignocellulosic biomass.

537828 Exploring Antimicrobial Properties of Lignin Derived Compounds and Materials

Abstract:

This study examines the antimicrobial properties of lignin containing hydrogel materials and thermochemically processed lignin-rich waste streams. We first explored the use of ionic liquids as solvents for creating hydrogels from mixtures of cellulose, xylan, Kraft lignin, in addition to both poplar wood and biomass sorghum to examine their effects on gel formation and antimicrobial properties against *Escherichia coli*. The chemical and physical properties of generated hydrogels were characterized using FT-IR, SEM, XRD, swelling kinetics, and stress-strain analysis. Additionally, two lignin rich streams, residues from the enzymatic hydrolysis of dilute acid and alkaline pretreated corn stover, were decomposed via pyrolysis and catalytic transfer hydrogenolysis, respectively. The resulting liquid bio-oils were subjected to sequential organic solvent extractions and characterized through HPLC and GC/MS. These extracted compounds were screened against *Saccharomyces cerevisiae* (yeast), *Escherichia coli*, and *Lactobacillus amylovorus* for antimicrobial properties. Six lignin model monomers: guaiacol, vanillin, vanillic acid, syringaldehyde, 2,6-dimethoxyphenol, and syringic acid were compared to the oils and extracted fractions to compare antimicrobial properties. This study provides insight into using ionic liquids for gel formation from lignocellulosic components and the antimicrobial potential of lignin derived compounds.

541565 25 by 25: Chemical Engineering in the Next 25 Years

Abstract:

In 2008, the centennial year for AIChE, we asked 25 chemical engineers for their visions of the profession from around the US and the world: Nine faculty, eight from industry, and eight post-docs and graduate students. We posed four questions related to each individual's industrial sector or area of research, asking for extrapolation, new impacts on existing sectors, new sectors, and an optional, more open-ended comment on the future of the profession as a whole.

In this 110th year of AIChE, we have revisited these questions with many of the same people. Ten years has seen many developments in our profession, in society, and in our thought-leaders' experiences. Here we report on their perspectives, comparing them to the 2008 visions.

541542 The Future of Chemical Engineering Itself

Abstract:

Chemical engineeringâ€™s future is bright because it is vital to so much of society and industry -- and its future is complicated by its rapidly expanding universe of activities. It must represent depth in core skills centered on the chemical sciences including biology and materials. At the same time, it must be transdisciplinary; it must exemplify the need and value of reaching across disciplines in borderless ways.

One way is to create and disseminate a compelling vision and strategy for chemical engineering, research, and education for the future. AIChE and the National Academies are working together to that end. In 1988, the National Academies published "Frontiers in Chemical Engineering: Research Needs and Opportunities.â€ Its committee of authors outlined a wide-ranging roadmap to turn promising research opportunities into reality, while calling for researchers to embrace new frontiers.

Since 1988, that has happened -- along with many other transformative changes like the Web, fracking, personalized medicine, and social media. A new nationwide effort, aided by international contributions, will address both the identity of chemical engineering and its future directions. Its goal is to ensure that the people in chemical engineering are primed to meet society's needs and rapid changes.

544708 Accelerating Innovation through Academic-Industrial Partnerships

Abstract:

In 2011, The Dow Chemical Company launched the University Partnership Initiative (UPI) to sponsor collaborative research at leading U.S. Universities. This 10-year commitment aims to address fundamental questions in industrial chemistry and chemical engineering with alignment to Dowâ€™s next-generation process technology and product development efforts. To date, this portfolio of projects has generated an impressive array of intellectual property, publications, and numerous examples of the collaborative developments implemented at Dow. This presentation will introduce UPI program and highlight several contributions from our collaborators and leaders in the chemical engineering profession.

540559 Maximizing Uptime, Efficiency, and Safety of Industrial Operations through Early Risk Detection

Abstract:

Achieving zero incidents while maximizing process uptime and efficiency is a key vision for operating companies. An important building block towards this vision is establishment of a proactive risk mitigation culture, supported by effective systems and a strong management. Studies have shown that most of the incidents and unexpected process failures can be avoided if the operating teams get timely information about developing risks and take preventive actions early on. While advances have been made in process monitoring and automation in the past decades, there still remains significant technological and behavioral gaps that prevent plant operations from proactively mitigating process risks.

A modern industrial plant monitors thousands of parameters, generating upwards of 50-100 million data points every day. Thanks to recent breakthroughs in machine learning and artificial intelligence approaches, today there are autonomous systems that can sift through this data and point out meaningful and timely insight. This can help operating teams ascertain process issues that are hidden in the data, long before process variables reach critical levels.

In this presentation, a new autonomous system, Dynamic Risk Analyzer™ (DRA), will be introduced that points out early indicators of process issues using proprietary machine learning. The risk indicators enable operations team to respond and make required process changes, days and sometimes weeks ahead of any alarm, avoiding late stage (expensive) fixes. To realize these objectives fully, a proactive management workflow must also be in place. The presentation will discuss new initiatives and workflows developed by plant operations to implement a proactive culture. Real-life case studies from plants using DRA technology will be presented on how this culture change resulted in increased uptime, efficiency and safety of plant operations.

531309 Gaussian Processes for Hybridizing Analytical & Data-Driven Decision-Making

Abstract:

Surrogate models are widely appreciated in process systems engineering [1]. The typical setting focuses on expensive-to-evaluate, possibly uncertain functions. Examples include: modular process simulators [2], integrated gasification combined cycle processes [3], a carbon capture absorber [4], and many other applications, e.g. [5-8]. Resources are typically limited, so effective decision making requires data-efficient learning.

The data science and statistical machine learning communities typically focus on models learned solely from observed data. But chemical engineering applications may also require explicit, parametric models, e.g. modeling known process constraints, operations constraints, and cost objectives [9]. So, work has integrated semi-algebraic functions with those learned from data [10] or developed semi-physical modeling techniques [11, 12].

This presentation surveys the state-of-the-art in hybridizing analytical and data-driven decision making [13]. We consider three probabilistic modeling applications to these hybrid situations:

Design of experiments for model discrimination [14]. We bridge the gap between classical, analytical methods [15] and Monte Carlo-based approaches [16]. Classical methods may have difficulty managing non-analytical model functions and data-driven Monte Carlo approaches come at a high computational cost. We replace the original, parametric models with probabilistic, non-parametric Gaussian process surrogates learned from model evaluations. The surrogates are flexible regression tools that extend classical analytical results to non-analytical models, while providing us with model prediction confidence bounds and avoiding the computational complexity of Monte-Carlo approaches.

Multi-objective optimization [17, 18]. We make novel extensions to Bayesian multi-objective optimization in the case of one analytical objective function and one black-box, i.e. simulation-based, objective function. The resulting method has been applied to a bone neotissue application [19] and a more general test suite.

Scheduling plant operations under uncertainty. For processes with equipment degradation, we use Gaussian processes to approximate large-scale, mixed-integer optimization problems.

We close by offering a broad outlook on applying probabilistic surrogate models to chemical engineering. Statistical machine learning has recently attracted significant interest in process systems engineering [20]. Here we show that state-of-the-art research in Gaussian processes [21, 22] and probabilistic modeling more generally [23] can have a big impact on chemical engineering.

References

[1] Bhosekar A, Ierapetritou M. Advances in surrogate based modeling, feasibility analysis and and optimization: A

review. *Comput Chem Eng*. 108: 250-267, **2018**.

[2] Caballero JA, Grossmann IE. An algorithm for the use of surrogate models in modular flowsheet optimization. *AIChE J*. 54(10):2633-2650, **2008**.

[3] Lang Y, Zitney SE, Biegler LT. Optimization of IGCC processes with reduced order CFD models. *Comput Chem Eng*. 35(9):1705-1717, **2011**.

[4] Cozad A, Sahinidis NV, Miller DC. Learning surrogate models for simulation-based optim-ization. *AIChE J*. 60(6):2211-2227, **2017**.

[5] Boukouvala F, Ierapetritou M. Surrogate-based optimization of expensive flowsheet modeling for continuous pharmaceutical manufacturing. *J Pharm Innov*. 8: 131, **2013**.

[6] Soepyan FB, Cremaschi S, Sarica C, and others. Estimation of percentiles using the Kriging method for uncertainty propagation. *Comput Chem Eng*. 93:143-59, **2016**.

[7] Shokry A, Ardakani MH, Escudero G, Graells M, Espuña A. Dynamic Kriging-based fault detection and diagnosis approach for nonlinear noisy dynamic processes. *Comput Chem Eng*. 106:758-76, **2017**.

[8] Tran AP, Georgakis C. On the estimation of high-dimensional surrogate models of steady-state of plant-wide processes characteristics. *Comput Chem Eng*. DOI 10.1016/j.comp-chemeng.2018.02.014, **2018**.

[9] Boukouvala F, Hasan MF, Floudas CA. Global optimization of general constrained grey-box models: new method and its application to constrained PDEs for pressure swing adsorption. *J Glob Optim*. 67(1-2): 3-42, **2017**.

[10] Boukouvala F, Floudas CA. ARGONAUT: AlgoRithms for Global Optimization of coNstrAined grey-box compUTational problems. *Optim Lett*. 11(5): 895-913, **2017**.

[11] Pearson RK, Pottmann M. Gray-box identification of block-oriented nonlinear models. *J Process Contr*. 10(4):301-315, **2000**.

[12] Cozad A, Sahinidis NV, Miller DC. A combined first-principles and data-driven approach to model building. *Comput Chem Eng*. 73: 116-27, **2015**.

[13] Boukouvala F, Misener R, Floudas CA. Global optimization advances in mixed-integer nonlinear programming, MINLP, and constrained derivative-free optimization, CDFO. *Eur J Oper Res*. 252: 701-727, **2016**.

[14] Olofsson S, Deisenroth MP, Misener R. Design of Experiments for Model Discrimination Hybridising Analytical and Data-Driven Approaches. *arXiv:1802.04170*. **2018**.

[15] Buzzi-Ferraris G, Forzatti P, Emig G, Hofmann H. Sequential experimental design for model discrimination in the case of multiple responses. *Chem Eng Sci*. 39(1): 81-85, **1984**.

[16] Vanlier J, Tiemann CA, Hilbers PAJ, van Riel NAW. Optimal experiment design for model selection in biochemical networks. *BMC Syst Biol*, 8(20), **2014**.

[17] Olofsson S, Mehrian M, Geris L, Calandra R, Deisenroth MP, Misener R. Bayesian Multi-Objective Optimisation of Neotissue Growth in a Perfusion Bioreactor Set-Up. *Computer Aided Chemical Engineering*. 40: 2155-2160, **2017**.

[18] Beykal B, Boukouvala F, Floudas CA, Pistikopoulos EN. Optimal design of energy systems using constrained grey-box multi-objective optimization. *Comput Chem Eng*. DOI 10.1016/j.compchemeng.2018.02.017, **2018**.

[19] Mehrian M, Guyot Y, Papantoniou I, Olofsson S, Sonnaert M, Misener R, Geris L. Maximizing neotissue growth kinetics in a perfusion bioreactor: An *in silico* strategy using model reduction and Bayesian optimization. *Biotechnol Bioeng*. 115(3):617-29, **2018**.

[20] Lee JH, Shin J, Realff MJ. Machine learning: Overview of the recent progresses and implications for the process systems engineering field. *Comput Chem Eng*. DOI 10.1016/j.compchemeng.2017.10.008, **2017**.

[21] Damianou A, Lawrence N. Deep Gaussian processes. In *Artificial Intelligence and Statistics*. 207-215, **2013**.

[22] Deisenroth MP, Ng JW. Distributed Gaussian processes. In *32nd International Conference on International Conference on Machine Learning*. 37: 1481-1490, **2015**.

[23] Ghahramani Z. Bayesian non-parametrics and the probabilistic approach to modelling. *Phil Trans R Soc A*. 371(1984):20110553, **2013**.

538530 Furan Production from Biomass Hydrolysates: Scale-up of a Novel, High-Yield "Sire" Process

Abstract:

Lignocellulosic biomass is a sustainable feedstock for production of furans – 5-hydroxymethy furfural (HMF) and furfural – versatile platform molecules that can be converted to drop-in fuels and precursors for synthetic materials. Lack of efficient lignocellulosic sugars-to-furan pathways (especially HMF) constitutes a major barrier to the industrial production of furans. We have recently demonstrated high yields of furans from biomass hydrolysates using a novel Simultaneous-Isomerization-Reactive-Extraction (SIRE) followed by Back-Extraction (BE) and dehydration process (*Green Chem.*, **2017**, 19, 1782). We are now in the process of developing a continuous-flow, meso-scale (5L) SIRE-BE system and this presentation will discuss furan production using concentrated AFEX (Ammonia Fiber Expansion) treated lignocellulosic biomass hydrolysate (60 g-total sugars/L). The continuous flow system consists of a packed bed Isomerization column containing immobilized glucose/xylose isomerase to convert aldose hydrolysate sugars (glucose and xylose) to more reactive keto-forms (fructose and xylulose). The isomerization column is closely coupled with a hollow fiber membrane module to facilitate continuous Reactive Extraction of the ketose sugars into an octanol phase. The aqueous- and organic- phases flow in the lumen- and shell- sides of a hydrophobic polypropylene module. Finally, ketose sugars Back Extracted from octanol into an immiscible acidic ionic liquid where they are dehydrated to furans. A key aspect of the scale-up is to develop and validate mathematical models that describe the mass transfer processes during Isomerization and Reactive Extraction. We will present data from continuous-flow studies and compare experimental results with model predictions. Techno-economic feasibility and sensitivity analysis will also be presented to assess the potential of the SIRE-BE process to provide low-cost furans for drop-in fuels and polymer precursors.

534284 A Novel and Green Approach in Engineering Transparent and Homogenous Cellulose Nanocrystal-Lignin UV Protection Films

Abstract:

A Novel and Green Approach in Engineering Transparent and Homogenous Cellulose Nanocrystal-Lignin UV Protection Films

In the context of valorization of lignin produced from the pulp and paper industry, a simple approach to prepare the lignin and cellulose nanocrystals (CNC) based transparent and homogenous UV protection films is developed. The results demonstrated for the first time that CNC aqueous suspensions with and without containing lignin could be tuned through the addition of sodium hydroxide (NaOH) to produce transparent and homogenous films. Initially, CNC films were optimized for improving their transparency using various NaOH additions. It was observed that the addition of NaOH in the range of 3 to 4 wt.% resulted in the CNC films with maximum transparency. Moreover, NaOH addition enhanced the homogeneity of the films by uniformly dispersing lignin in the films. CNC/lignin films were prepared by using optimized CNC suspensions with lignin concentrations in the range 1 to 10 wt.%. CNC/AL (alkaline lignin) and CNC/SKL (softwood kraft lignin) with 10 wt.% lignin concentration provided complete UV blocking. The UV protection behavior of these films was stable under UV irradiation. Cross polarized optical microscopy and scanning electron microscopic images of films showed some degree of global alignment of CNC rods upon addition of NaOH, which remained unaffected by lignin addition. Acetylation of lignin reduced the lignin color with only slight reduction in their extinction coefficient. Incorporation of 10 wt.% Ac-SKL into CNC films increased visible light transmittance at 550 nm by 67% without significantly affecting the UV blocking property compared to 10 wt.% CNC/SKL films. Presence of lignin also provided the thermal and contact angle stability. 10 wt.% SKL addition increased the maximum weight loss temperature (T_1) of CNC by 5 °C, while 10 wt.% Ac-SKL, being more stable, increased T_1 by 20 °C. Due to the relatively more hydrophobic nature of lignin compared to CNC, contact angle of CNC/lignin films remained stable over time compared to CNC films; however, initial contact angle was unaffected by the presence of lignin. With further optimization in terms of lignin color reduction and increasing the CNC alignment, these CNC/lignin nanocomposites could be a potential biodegradable, low cost alternative as a coating/film material with UV blocking and optical polarization functionality for sunglasses, automobile windshields, home windows, contact lenses and UV sensitive polymers.

534378 Effect of Activation Temperature on Oxygen Functional Groups and Corresponding Electron Exchange Capacities on Hydrochar

Abstract:

Hydrothermal carbonization (HTC) a thermochemical process, where biomass is treated with high-pressure high-temperature liquid water. Hydrochar, the solid product of HTC, are carbon-dense microspheres and contains oxygen functional groups on the surface. Among other process parameters, HTC temperature is the most dominant for hydrochar functional properties.

The surface functional groups on the surface of chars play a vital role to use them as an adsorbent. Some research has been done some functional group analysis for hydrochar at different HTC temperature. However, this analysis is still required for the activated hydrochar. Although, the surface area and porosity of hydrochars increase with activation temperature, it is imperative to understand the content of such groups on the surface of activated hydrochar for its multifunctional applications.

Therefore, the objectives of this study were to quantification of oxygen functional groups and electron exchange capacities on activated hydrochars from model compounds (cellulose, wood). The hydrochars were activated in a Muffle Furnace at three different temperatures (400 °C, 500 °C, and 600 °C) for 1 hour. The change in functional groups on activated hydrochar surface was evaluated by surface morphology analysis, pH, Boehm titration, PZC, and FTIR. As prepared hydrochar shows relatively low surface area, mostly because of the lack of pores/ clogged pores with volatiles. An increase of surface area was observed but the oxygen functional groups were decreased with the increase of activation temperature as evident by the increase of pH, pKa, PZC. EAC values were in the range of 1.7-1.83 mmol e-/g activated hydrochar

541026 Sustainable Food Industry Wastes for Biocomposite Applications

Abstract:

Fruit wastes are a major byproduct of the food and wine processing industries. These byproducts, containing the skin, seeds, and stems of the fruit are left over after processing and are typically sent to landfills. Apple pomace (AP) and grape pomace (GP) are two examples of such byproducts that can be utilized as low-cost filler for plastic composite applications. Once dried, the pomace is ground and incorporated into a polymer matrix, with as much as 50 % by weight pomace. This addition reduces the amount of pomace sent to the landfill, increases its value and decreases the amount of expensive polymers required. The pomace can be combined with biobased polymers to create green composites. Biobased poly(butylene succinate) (BioPBS) is an expensive biobased, biodegradable polyester whose cost can be reduced with the incorporation of a filler. AP and BioPBS composites and GP and BioPBS composites were created and mechanical and thermal analysis of the composites shows comparable properties to neat BioPBS. The creation of sustainable biocomposites from a traditionally thought of waste product provides an environmentally friendly solution to a common processing problem, while simultaneously providing cost savings to plastic manufactures.

Acknowledgements

This research is financially supported by the Ontario Ministry of Agriculture, Food and Rural Affairs (OMAFRA), University of Guelph, Bioeconomy Industrial Uses Research Program Theme Project # 030177; and the Natural Sciences and Engineering Research Council (NSERC) Grants Project #400320, Canada Discovery Grants Project # 400320. Special thanks to Martinâ€™s Processing Ltd for the apple pomace samples and Andrew Peller Winery for the grape pomace samples.

548791 Sustainable Biomass Derived Fluorescent Carbon Dots for the Selective Detection of Fe^{3+} Ions

Abstract:

In this report, an economic and green one-step heating process was used to synthesize fluorescent carbon dots (CDs) and doped CDs using *Miscanthus* grass as the starting material. The structural characterization of as-synthesized *Miscanthus* derived CDs (MCD) and doped MCD was carried out by microscopic and spectroscopic study. The presence of copious oxygen containing functional groups was confirmed by energy dispersive X-ray spectroscopy (EDS) and Fourier transform infrared (FTIR). The synthesized MCD and doped MCD demonstrated high quantum yield. Photoluminescence (PL) and UV-visible spectroscopy was utilized to investigate the optical properties of MCD and doped MCDs. Their optical performance was further investigated under different pH and also their capacity to sense different metal ions were evaluated. N-doped MCD exhibited superb selectivity toward Fe^{3+} ions, with a low limit of detection and wide detection range. The linear relationship between normalized fluorescence intensity of the prepared N-doped MCD and the Fe^{3+} ions concentration can be utilized for selective, sensitive and label-free detection of Fe^{3+} ions.

Acknowledgements:

The authors would like to thank the following for their financial support: Ontario Ministry of Agriculture, Food and Rural Affairs (OMAFRA)/University of Guelph - Bioeconomy for Industrial Uses Research Program (Project # 030332); the Natural Sciences and Engineering Research Council (NSERC) Grants Project #400320, Canada Discovery Grants Project # 400320 and 401111; and the Ontario Research Fund, Research Excellence Program; Round-7 (ORF-RE07) from the Ontario Ministry of Research, Innovation and Science (MRIS) (Project # 052644 and 052665).

533729 Oxidative Pretreatment Process of Sugarcane Bagasse Assisted By Hydrodynamic Cavitation

Abstract:

Hydrodynamic cavitation (HC) has recently been reported as a new promising pretreatment technology for lignocellulosic substrates with an added advantage of simplicity and easy scalability. On the other hand, oxidative pretreatment is another pretreatment which require a long process time and high dose of chemicals for this process to be declared as an efficient one. Therefore, In this work HC was employed to improve the potassium persulfate/hydrogen peroxide-alkaline medium pretreatment. At the optimum pretreatment conditions, about 32 g/L and 16 g/L of glucose and xylose were achieved in the enzymatic hydrolysis process with 10% of solid/liquid ratio using the SCB pretreated under the following conditions: 1% of H_2O_2, 0.2 of NaOH and 25 mmol of $K_2S_2O_8$, 10 min, 70 â °C and 3bar of inlet pressure. Our initial result is quite promising as compared with the previous ones using alkaline hydrogen peroxide assisted by HC for sugarcane bagasse.

526555 Catalytic Etherification of Glycerol to Glycerol Oligomers in the Presence of Alumina Supported Ca/Sr Mixed Oxides

Abstract:

In this work, the solvent-less etherification of glycerol to oligomers using alumina supported Ca and Sr catalysts was studied. These catalysts were obtained with impregnation of strontium and calcium onto γ-alumina (γ-Al_2O_3) in $Sr(NO_3)_2$ and $Ca(NO_3)_2$ solutions. The evidence for synergistic effects between active Ca and Sr species was demonstrated by etherification tests over a series of Ca-Sr-Al catalysts with varied Ca-to-Sr molar ratios. Effect of calcination temperatures of these supported catalysts, from 600 °C to 900 °C, on the resultant reusability and conversion efficiency of glycerol to oligomers and the selectivity of linear and branched oligomer to cyclic ones were investigated. The catalyst properties were characterized by various instruments including BET, TPD, XRD and SEM, and the ICP-OES analysis was performed to confirm the leaching of the metal species in the liquid phase of the reaction mixture. With Ca and Sr supported on γ-alumina catalyst, the catalyzed etherification reactions were carried out at 250 °C with 2.5 wt% Ca/Sr/γ-alumina calcined at 900 °C in nitrogen atmosphere. Under optimal impregnation conditions, a conversion efficiency of glycerol 80 % glycerol could be obtained after 10-h reaction. Moreover, a selectivity near 70% of di- and tri-glycerol in linear and branched forms over the cyclic form could be achieved.

538721 The Role of Copper Stability in Selectively Condensation of Ethanol to Higher Alcohols

Abstract:

The recent emergence of a strong bioethanol industry has elevated interest towards its use as a sustainable platform molecule towards producing high valued chemicals; one such process is the selective conversion of ethanol into high value C_{4+} alcohols. This multistep reaction involving acid-base catalyzed condensation and hydrogen transfer tends to yield a complex mixture of side products such ethers, esters, and ketones that prevent efficient conversion. Our work has demonstrated 80% selectivity to higher alcohols with conversion above on MgAl hydrotalcite catalysts at low copper loadings, with stable catalyst lifetimes over 100 hours in a flow through-reactor setup. Lower concentrations of copper was shown to prevent the formation of ester and ketone side products without reducing the conversion of ethanol. Characterization of the catalyst by STEM and XAS EXAFS work reveal that this is largely due to high dispersion of copper at lower copper loadings with near absence of copper clustering, even under high temperatures and reductive conditions. Further *in operando* XANES and EXAFS study shows that copper only reduces partially to Cu

support matrix allow for the production of dispersed Cu^{+1} sites, which promotes the efficient catalytic reaction to higher alcohols while not contributing to side reaction pathways such as esterification and C-C scission. The dispersion and stability of these individual copper promoters is critical to the high efficiency of this reaction. This work will discuss the limiting copper movement within the support to lengthen the lifetime of catalyst activity as well as throughout multiple regeneration cycles.

538786 Effects of Alloying Pd and Cu on Tandem Dehydrogenation-Aldol Condensation Reactions

Abstract:

The successful reduction of carbon dioxide emissions as a result of the production and consumption of fuels and chemicals requires the use of biomass resources as a feedstock. As a result of this, the discovery of novel pathways to upgrade platform molecules to fuel precursors is of critical importance. More specifically, for the production of diesel substitutes, the formation of new C-C bonds is required, as platform molecules deriving from the acid-catalyzed hydrolysis of biomass or the fermentation thereof are in the C2-C6 carbon range.

In this work, we investigate the upgrade of fermentation-derived acetone-butanol-ethanol mixtures to renewable diesel precursors. To achieve this, we develop a bimetallic Pd-Cu catalyst. We show that alloying of Pd and Cu results in the suppression of undesirable decarbonylation reactions. We present X-ray absorption spectroscopy results showing the segregation of Cu to the catalyst nanoparticle surface. Kinetic measurements under flow conditions demonstrate that Pd-Cu surface ensembles are active for hydrogenation reactions, while Pd-Pd ensembles catalyze C-C bond scission reactions. Density functional theory and microkinetic modeling show that the PdCu ensembles are bind CO fragments weakly, and hence destabilize C-C bond scission transition states. Similarly, the weaker CO binding onto low-index planes compared to stepped surfaces results in higher decarbonylation rates over smaller nanoparticles. Based on these results, we generalize our conclusions for a range of reactions, under oxidizing and reducing conditions, and present the implications of the segregation of alloy components to the catalyst surface on reactivity and selectivity.

References:

1. *Nature* 491 (2013) 235â€"239
2. *Am. Chem. Soc.* 138 (2016) 6805-6812
3. *Catal. Sci. Technol.* 8 (2018) 314-327

530475 Lubricant Base Oils Production from Biomass

Abstract:

The gradual depletion of world fossil fuel resources and global environmental concerns have prompted the use of biomass feedstock for renewable products. Lubricant is one such product that is primarily obtained from petroleum and it is introduced to reduce friction between surfaces in mutual contact. Base oils are the main components of lubricant and the majority of them are petroleum-derived mineral oils and synthetic oils (e.g. poly-Î±-olefins). The oils are formulated with different additives into several grades to suit automotive and industrial applications. Production of lubricant base oils from biomass will be an attractive alternative to petroleum derived-based oils to improve sustainability. So far, a few efforts have been devoted to the synthesis of lubricant-based oil alicyclic hydrocarbons, from biomass furans.[1-2] In this work, we report a novel route for selective production of base oil hydrocarbons from biomass platform chemicals and natural oils. The overall yield of hydrocarbons could reach up to 80%. Importantly, the hydrocarbon structures can be tailored using different biomass derived compounds to tune properties for various applications.

1. Balakrishnan, E. R. Sacia, S. Sreekumar, G. Gunbas, A. A. Gokhale, C. D. Scown, F. D. Toste and A. T. Bell, Proc. Natl. Acad. Sci. USA 2015, 112, 7645-7649.
2. Balakrishnan, G. E. Arab, O. B. Kunbargi, A. A. Gokhale, A. M. Grippo, F. D. Toste and A. T. Bell, Green Chem. 2016, 18, 3577-3581.

530529 Selectivity Control during the One-Pot Conversion of Aliphatic Carboxylic Acids to Linear Olefins through Tandem Hydrogenation/Dehydration

Abstract:

Olefins are important building blocks for the synthesis of higher value chemicals in the petrochemical industry and they are typically produced from fossil fuels. As an alternative to fossil fuel-based chemicals, biomass-derived carboxylic acids could serve as a viable replacement feed for olefin production. However, the production of olefins from carboxylic acids in one step remains a challenge due to the stability of carboxylic acids and the tendency of olefins to over-hydrogenate during hydrogenation, leading to alkanes.

In the present study, we have studied the one-pot catalytic conversion of linear aliphatic carboxylic acids to linear olefins via tandem hydrogenation/dehydration reactions. Hexanoic acid was converted to a mixture of hexenes over Cu nanoparticles supported on commercial silica-alumina (Siral 70) in a continuous flow reactor. At 483 K and 5 bar H_2 gauge pressure, we obtained 98.5% conversion of hexanoic acid, with a molar product distribution of 88.9% hexenes and 11.1% hexane. Remarkably, we observed a brusque selectivity switch to 99.9% hexane at 100% conversion.

A product intermediates study shows that when switching from a feed of pure hexene to a mixture of hexene and hexanoic acid, overhydrogenation of hexene can be suppressed. In addition, hexene temperature-programmed desorption (TPD) and FT-IR spectroscopy reveals that adsorption of hexanoic acid prevents hexene from accessing the catalyst surface, thus avoiding overhydrogenation. Reactions using butyric acid as feed also exhibit similar phenomena. We have also performed reactions utilizing butyric acid derived from biomass fermentation, albeit with significant catalyst deactivation.

We propose that the presence of a small amount of carboxylic acid on the surface of the catalyst prevents overhydrogenation of the olefins. In light of these results, current efforts are targeted at applying this concept to explain similar phenomena in other reactions, particularly in biomass conversion. The presence of numerous carboxyl groups in native/pre-treated biomass might explain the difference in catalyst performance/selectivity when processing model feed versus real biomass. Furthermore, the addition of carboxylic acids to the reaction medium during biomass upgrading could afford more control over the selectivity of the reaction.

526827 Mechanism and Kinetics of Isobutene Production over Zirconia-Supported Zinc Oxides

Abstract:

Growing interest in utilizing biomass-derived platform chemicals to produce fuels and specialty chemicals has motivated the investigation of the reaction of ethanol or acetone to form isobutene, a valuable specialty chemical used in the production of fuel additives, polymers, and other high-value products. Reports in the literature have identified zinc-zirconium mixed oxides as effective and selective catalysts for the production of isobutene from ethanol at 723 K. In this investigation, zinc-zirconium mixed oxides prepared via incipient wetness impregnation were selective and highly active catalysts for the acetone to isobutene reaction, reaching near-theoretical isobutene selectivities of over 86%. The mechanism and kinetics of the acetone and ethanol to isobutene reactions were studied in order to elucidate the reaction pathway, the roles of active acidic and basic sites, and the role of water in promoting stability and selectivity. Zinc-zirconium mixed oxide catalysts with varying zinc loadings were synthesized and characterized with XRD, BET, CO2-TPD, and DRIFTS FTIR with pyridine adsorption. Kinetic studies suggested that the ethanol to isobutene reaction proceeds via dehydrogenation to acetaldehyde, followed by condensation to form acetic acid, aldol condensation to produce acetone, dimerization of acetone to form diacetone alcohol, dehydration to form mesityl oxide, then finally hydrolysis of mesityl oxide to produce isobutene and acetic acid, which undergoes further aldol condensation to produce acetone. Understanding the reaction pathway, mechanism, and roles of the acidic and basic sites can help to define the catalyst properties required to selectively produce isobutene from a variety of sustainably-derived platform molecules.

533068 Catalytic Upgrading of Sugar-Derived Polyols to Glycols in Absence of Externally Added Hydrogen

Abstract:

Due to the increasing demand for fossil fuels and their diminishing supply worldwide, it is of considerable importance to utilize renewable resources for sustainable development of human society. Replacing fossil-based feedstock with biomass-derived ones to produce renewable fuels and chemicals is one of the promising ways to address challenges facing the energy and chemical industries. Sugar polyols, such as sorbitol, xylitol and erythritol are attractive platform chemicals because they can be readily derived from cellulosic biomass. Hydrogenolysis of sugar polyols can be converted to a variety of industrially important chemical intermediates including ethylene glycol (EG) and 1,2-propanediol (1,2-PDO), which are widely used as antifreezes, solvents and pharmaceuticals.

Up to date, it is known most existing hydrogenolysis processes are carried out under high temperature (T > 220 °C) and high H_2 pressure conditions (P > 4 MPa), which cause several technological issues such as poor product selectivity due to significant side reactions, fast catalyst deactivation as well as high capital cost and energy consumption.

In this work, we proposed a novel one pot H_2 generation and hydrogenolysis process in absence of external H_2 under mild conditions (T: 160 °C-200 °C, P: 1.0 Mpa nitrogen atmosphere).We have designed a series of bimetallic Pd-M (M: Cu, Ni, Pt, *etc*) catalysts supported on active carbon and layered double hydroxides and confirmed their remarkable performances in catalytic conversion of xylitol, sorbitol and mannitol to 1,2-PDO, EG and lactic acid (LA). In particular, catalyst activity and selectivity will be studied for different catalysts. Structure-performance correlations for H_2 generation and hydrogenolysis will be established based on bulk and surface characterization using XRD, SEM, TEM and chemisorption. The bimetallic Pd-Ni/AC and Pd-Pt/AC catalysts showed superior hydrogenated activities as well as selectivities towards lactic acid and glycols, compared with other bimetallic catalysts under identical conditions. At nearly 60.0% sorbitol conversion, a selectivity of 36.5% to ethylene glycol, propanediol and other glycols was obtained on Pd-Ni/AC at 200 °C, 1.0Mpa nitrogen atmosphere in absence of external H_2. Here we also researched reaction mechanism or pathways by studying metal active sites, H_2 spillover effect and the role of solid-base via regulating the growth of Pd-M and the ratio of acidity (basicity). It was concluded that H_2-spillover from Pd to M and solid-base acidity (basicity) may strongly influence the reaction mechanism to some extent, thereby influence the activity and selectivity of hydrogenolysis of biomass-derived to chemicals. Further details on reaction pathways and kinetic modeling will also be discussed in this presentation.

533584 Selective C-C Bond Scission of Primary Alcohols Using Cerium Oxide-Supported Palladium Catalyst

Abstract:

Biomass attracts significant attention as a carbon-neutral resource with the potential to suppress the CO_2 emissions. Recently, the biomass-derived chemicals such as glycerol, furfural, and levulinic acid are identified as the best suited alternatives for petroleum-derived chemicals. From the stand point of green sustainable chemistry, development of highly selective catalytic transformation of these compounds to valuable chemicals has been desired. To date, much effort has been devoted to selective C–O bond cleavage of high-oxygen containing biogenic polyols, such as glycerol, to produce valuable chemicals by hydrogenolysis and deoxydehydration.[1,2] On the other hand, C–C bond cleavage has not yet been widely researched despite its great potential to extend the utility of biomass-derived chemicals. The reported C–C bond cleavage methods often suffer from low selectivity toward the desired chemicals and limited substrate scope. In our group, we have reported the selective C–C bond scission of platform biomass-derived chemicals catalyzed by Ru/CeO_2 in the presence of water under 3 MPa of H_2.[3] In order to open new routes for industrially important chemicals from a wide range of biomass derivatives, development of highly selective and greener C–C bond scission catalysts is desired.

In this work, we found that cerium oxide-supported palladium (Pd/CeO_2) efficiently promoted the selective one-carbon scission of primary alcohols under air without H_2. The cleavage of C–C bonds occurs at positions adjacent to the hydroxymethyl groups (Scheme 1). The solid catalyst was reusable without appreciable loss of the activity and selectivity.

References

[1] T. Mizugaki, Y. Nagatsu, K. Togo, Z. Maeno, T. Mitsudome, K. Jitsukawa, K. Kaneda, *Green Chem.* **2015**, *17*, 5136–5139.

[2] T. Mizugaki, K. Togo, Z. Maeno, T. Mitsudome, K. Jitsukawa, K. Kaneda, *ACS Sustainable Chem. Eng.* **2016**, *4*, 682–685.

[3] T. Mizugaki, K. Togo, Z. Maeno, T. Mitsudome, K. Jitsukawa, K. Kaneda, *Sci. Rep.* **2017**, *7*, 14007.

524736 Gas Phase Catalytic Oxidation of Lignin to Produce Phenolic Compounds over Vanadia Catalysts

Abstract:

The pretreatment of lignocellulosic biomass to create fermentable sugar streams *en route* to fuel/chemical production may also produce streams of insoluble lignin. These lignin stream can be burned to produce process heat and power. An alternative use of the lignin streams could be the production of valuable products such as phenolic compounds. This work explores the pyrolysis of lignin residue to produce gas phase products which can be passed of vanadia-based catalysts to oxidatively remove methoxy side chains to produce phenolic compounds, which could be used in the production of polymers and resins. Because lignin contains a complex mixture of linkages with methoxy groups, as well as both C-C and C-O bonds linking phenyl groups, there are number of bonds that must be broken in order to funnel lignin into the desired slate of phenolic compounds without methoxy groups (simple phenols). The subject of this presentation uses pyrolysis of lignin to create a mixture of gas phase species that can be passed over a catalyst to oxidatively cleave the methoxy groups to produce simple phenols. Because the goal is to partially oxidize the lignin pyrolysis vapors to produce phenols, as opposed to over-oxidation which would result in CO and CO_2 production, the oxidation activity of the catalysts must be controlled. Catalysts with varying vanadium-oxygen bond strength, which affects the oxidation activity, were designed by altering the catalytic support and synthesized using incipient wetness impregnation. Reaction experiments were conducted by passing lignin pyrolysis vapors over the catalysts at 500 °C and were shown to produce simple phenols from methoxy phenols. The effect of catalyst composition will be discussed along with relevant characterization techniques, coupled with computational modeling, to help understand the role of metal-oxygen bond strength in the catalytic activity.

526393 Mechanistic Study of the Hydrogenolysis of Diaryl Ethers Catalyzed By Heterogeneous Metal Catalysts

Abstract:

Manipulating reaction pathways of metal-catalyzed reactions is key to efficiently convert polar aromatic compounds. Diaryl ethers undergo C-O hydrogenolysis and hydrogenation with rates that depend on the specificity of the metal. We are studying the corresponding catalytic chemistry of Ni, Pd, Ru and Pt with reaction kinetics and experiments with isotopically labeled reactants. We have found a common reaction mechanisms for all studied metals but different isotopic effects, which points to the interaction of hydrogen with the metal as the factor determining selectivity. The details of the mechanisms, which are being tested by theoretical calculations, will be reported and selectivity rationalized as a function of the binding energy of reactants with the metal.

527899 Palladium-Iron Bimetallic Catalyst: High Activity and Stability for Aqueous Phase Hydrogenations

Abstract:

A novel palladium-iron bimetallic catalyst was synthesized that provided high hydrogenation activity and stability relative to palladium only under harsh hydrothermal conditions. To generate the catalyst, palladium was dispersed on carbon coated silica, which was doped with iron. The catalyst was applied to the hydrogenation of carbonyl groups in acetone, 2-pentanone, and propionaldehyde. The Pd-Fe catalyst gave more than a three-fold higher hydrogenation activity than the nanoparticle Pd only counterpart, while Fe nanoparticles were shown to be inert. The synthesis procedure was optimized for the bimetallic material, which demonstrated that the introduction of Fe contributed to a higher turnover frequency rather than simply increases the number of active sites. Interestingly, the same synthesis procedure was used to synthesize a Pd-Mo bimetallic catalyst that also increased activity.

536491 Hydrodeoxygenation of Guaiacol over Ni and Mo Nanoparticles Supported on SBA-15 and γ-Al_2O_3.

Abstract:

1. Introduction

The limitation of fossil energy resources and the increasing need for chemicals and liquid fuels require strong efforts in utilizing renewable energy sources. In this way, a technology such as biomass to liquid fuels (BTL) is being focus. The fast pyrolysis is a promising tool to transform various biomass feedstock's, lignocellulosic biomass, into an energy-dense and oily mixture. However, it possesses considerable amounts of oxygenates molecules, phenolic compounds and water in the material obtained by this method make it impossible to use the final product as motor fuel. For the use in chemical industry or fuel production an upgrading of the pyrolysis oil is necessary, which can be achieved via hydrodeoxygenation (HDO). This process consists in the treatment of bio-oil at temperatures between 200 and 450 °C with hydrogen at high pressure (10–300 bar) in the presence of a catalyst, so that the oxygen contained in organic molecules can be extracted with water formation[1]. Research efforts to study the chemistry of hydrodeoygenation of biomass-derived oil are gaining considerable importace over the last 20 years[1].The most frequently examined HDO catalysts commercially used for petroleum hydrotreatment products are NiMo/Al_2O_3 and CoMo/Al_2O_3.In the present study catalysts based on supported Ni and Mo are evaluated as catalysts for HDO reaction. The effect of metal phase and support properties, SBA-15 and γ-Al_2O_3 (reference support), as well as their interactions has been investigated. For the catalytic reactions, guaiacol was chosen as a model compound for lignin depolymerization fragments formed during bio-oil pyrolysis from lignocellulosic biomass.

2. Methods

The pure mesoporous silica SBA-15 was synthesized by hydrothermal method. The γ-Al_2O_3 was purchase from Degussa. The bimetallic nickel-molybdenum catalysts were prepared by wet impregnation. First, appropriate quantities of $(NH_4)_6Mo_7O_{24}.4H_2O$ and $Ni(NO_3)_2.6H_2O$ were dissolved in deionized water; the support was added and stirred for 2 h. Then, the suspension was freeze dried and the catalyst was calcined at 500 C for 2 h, under air flow. The metal contents were 15% of MoO_3 and 5% de NiO (weight percentage) with relation to support. The materials were characterized by nitrogen physisorption isotherms, transmission electron microscopy (TEM), X-ray powder diffraction (XRD), energy-dispersive spectroscopy (EDS) and with Raman and infrared spectroscopy. The total acidity of the catalysts was measured by Temperature programmed desorption of NH_3 (TPD-NH_3). The catalytic tests were carried out in a fixed bed flow reactor system under 15 kgf/cm^2 pressure. The catalysis, mass around 200 mg, was pre-treated in situ with 50 cm^3/min flow of pure H_2 at 500° C for 4 h and then cooled to the reaction temperature (200, 250 or 300° C). The feed mixture consisted of 2% w/w guaiacol (MERCK) in heptane P.A. (Aldrich) with 14.8

cm^3/h flow and 100 cm^3/min H_2 flow. The feed and the products were analyzed by gas chromatography Shimadzu CG17A with FID detector, capillary column CPsil 5CB and automatic injection.

3. Results and discussion

The fresh catalysts X-Ray diffractogram show that the Ni and Mo were successfully incorporated under the SBA-15 and Al_2O_3 supports as NiO and MoO_3 nanoparticles. After nickel and molybdenum incorporation the surface area of the SBA-15 decrease from 630 to 332 $m^2.g^{-1}$. The same behavior was observed for γ-Al_2O_3 support, the surface area chance from 279 to 233 $m^2.g^{-1}$. The temperature reduction of Ni and Mo oxide were determined by TPR-H_2, 500 °C under H_2 atmosphere for 4 hours. The catalysts total acidity for NiMo/SBA and NiMo/ALU were 470 and 548 $\mu mol.g^{-1}$, respectively. The results of catalytic activity for both materials are summarized in Table 1. For NiMo/SBA and NiMo/ALU the reaction temperature have a positive effect in guaiacol conversion, the higher results were observed at 300 °C 84.7% and 72.6% for NiMo/SBA and NiMo/ALU, respectively. However, the higher HDO values occurred for reaction temperature at 200°C. Thus, the selectivity for deoxygenated compounds were more pronounced in low temperatures for both catalysts. The principal products formed during the guaiacol HDO reaction were benzene, cyclohexane, cyclohexene, phenol and o-cresol. Compared to NiMo/ALU the NiMo/SBA showed the best performance in HDO, HDA and higher selectivity for benzene in all reaction temperature studied.

Table 1: The catalytic results of NiMo/SBA and NiMo/ALU in the hydrodeoxygenation of guaiacol.

	NiMo/SBA			NiMo/ALU		
Temperature (°C)	**200**	**250**	**300**	**200**	**250**	**300**
Conversion (%)	3.7	74.5	84.7	5.1	45.0	72.6
HDO (%)	92.1	74.3	66.5	37.9	31.5	30.8
HDA (%)	84.2	48.7	35.3	23.1	12.3	2.7
Selectivity (%)						
Benzene	16.7	8.6	1.3	3.3	1.1	0.3
Cyclohexane	29.1	13.9	11.0	6.8	3.3	0.4
Cyclohexene	38.5	26.2	19.5	12.9	7.9	1.9
Phenol	10.5	17.4	34.2	16.3	24.8	41.3
O-Cresol	0.0	18.4	8.0	7.7	8.5	11.3

With the catalysts thermogravimetric analysis were possible quantified the coke formation after HDO reactions. For NiMo/ALU and NiMo/SBA the coke formation were 1,65 and 1.14 $mg.g_{cat}^{-1}.h^{-1}$.

4. Conclusions

Compared with NiMo/ALU the NiMo/SBA show the best performance in guaicol conversion, selectivity for deoxigenated compound and in HDO percentage. The principal compounds formed during the HDO reaction with NiMo/SBA were cyclohexene, cyclohexane and bezene. For NiMo/ALU the principal compounds were phenol and o-cresol. Thus, the mesoporous silicon oxide (SBA-15) was the best support for nickel e molybdenum nanoparticles resulting in the catalyst for guaicol hydrodeoxygenation.

References

- Arun, R.V. Sharma, A.K. Dalai, Renew. Energy Rev., 48 (2015) 240-255.

Keywords

Hydrodeoxygenation, catalysis, nickel, molybdenum.

530385 Hydroprocessing of Biomass-Derived Oxygenates on Metal-Exchanged Zeolites Using Light Alkanes As the Source of Hydrogen

Abstract:

The broader use of natural gas and biomass for energy applications is limited because of the short hydrocarbon chains and/or high oxygen content of the molecules that comprise these resources. Current technologies for natural gas and biomass upgrading require high temperatures and suffer from poor selectivities. Furthermore, biomass deoxygenation requires external sources of H_2 molecules that are most commonly derived by sacrificing a portion of the feedstock in steam reforming reactions. In the absence of molecular hydrogen, a portion of the carbon feedstock is converted to arenes during hydrogen transfer reactions to satisfy the stoichiometry for oxygen removal as water. This work aims to improve selectivity and solve the hydrogen deficiency issue by developing new catalysts that can facilitate hydro-deoxygenation reactions using light alkanes as the hydrogen source. These hydrogen transfer steps have the potential to selectively terminate growing hydrocarbon chains via desorption as less reactive alkanes, thereby producing targeted classes of compounds, as opposed to a wide range of products over a broad carbon number distribution. Furthermore, this strategy has the potential to upgrade a portion of the light alkane feed because the hydrogen transfer steps generate alkenes that can participate in subsequent C-C bond formation reactions.

Reactions of isobutane and n-butanal (either as single reactants or co-feeds) on H-BEA zeolite and H-BEA samples partially exchanged with Zn cations were studied to probe the effects of alkanes as co-reactants during acid-catalyzed reaction of oxygenates. The Zn-exchanged BEA zeolite was synthesized via ion-exchange using aqueous solutions of zinc nitrate followed by treatment in flowing, dry air first at 373 K for 20 h then at 773 K for 4 h. This method has been previously shown to effectively exchange protons for Zn^{2+} ions within the cages of zeolites. n-Butanal reactions (2 kPa) on H-BEA predominantly form 2-ethyl-2-hexenal via aldol-condensation (60% selectivity to the aldol-condensation dimer) while also forming xylenes and cyclic ketones via cyclization/deoxygenation of 2-ethyl-2-hexenal (25% selectivity). n-Butene formation was also observed (selectivity of 15%) via hydro-deoxygenation of n-butanal. Addition of isobutane as a co-feed on H-BEA leads to minor changes in the product selectivity, indicating that the unsaturated species formed from n-butanal conversion are not strong enough hydride acceptors to activate isobutane via hydride transfer on Bronsted acid sites. Product selectivities for n-butanal conversion on Zn-H-BEA were similar to those on H-BEA, suggesting a minor effect of the Zn sites on n-butanal reaction paths. Co-reaction of isobutane (40 kPa) and n-butanal (2 kPa) on Zn-H-BEA exhibited a completely different product distribution. The predominant products were C_4 hydrocarbons (34% selectivity with 50% selectivity to n-butane within C_4) and C_8 alkanes (dimethyl-hexane and methyl-heptane isomers; 31% selectivity). n-Butane is the expected isomer formed from deoxygenation-hydrogenation of n-butanal, and the C_8 isomers that were produced were those that would be expected from deoxygenation of aldol-condensation (methyl-heptanes) and Prins-condensation (dimethyl-hexanes) products. Reaction of isobutane/n-butanal feeds on the Zn-H-BEA catalyst also exhibited lower selectivity to the aldol-condensation dimer (16% versus 70%) and to cyclic species (16% versus 25%) compared to isobutane/butanal reactions on H-BEA. Isopentane formation (5% selectivity) and C_7 alkanes (methyl-hexanes and n-heptanes; 9% selectivity) were also observed on Zn-H-BEA during conversion of isobutane/butanal feeds. These C_7 isomers are what would be expected from deoxygenation of C_7 ketones derived from Tishchenko esterification-ketonization routes. Reaction of isobutane (40 kPa) was also carried out on H-BEA and Zn-H-BEA to identify the extent of isobutane dehydro-oligomerization reaction paths. Isobutane conversion was only 0.1% on H-BEA and only 0.2% on Zn-H-BEA, indicating that dehydro-oligomerization of isobutane only occurs to a negligible extent in parallel with n-butanal conversion paths. This negligible conversion of isobutane is consistent with the absence of trimethyl-pentane isomers from the product distributions (indicating undetectable isobutane dehydro-oligomerization).

530346 Flowthrough Reductive Catalytic Fractionation of Biomass

Abstract:

As an abundant and renewable source of carbon, lignocellulosic biomass is an excellent feedstock for sustainable fuel and chemical production. Biomass primarily consists of carbohydrates (cellulose and hemicellulose) and lignin. Carbohydrates have been extensively studied and shown to be versatile feedstocks for the production of fuels and chemicals. Lignin, the largest source of natural aromatics, becomes highly recalcitrant during carbohydrate processing, hindering chemical production from lignin. Reductive catalytic fractionation (RCF) has emerged as a leading biomass fractionation and lignin valorization strategy that simultaneously extracts, depolymerizes and stabilizes lignin without disturbing the carbohydrate fractions. Here, flowthrough reactors were used to investigate RCF of poplar. Most RCF studies to date have been conducted in batch, but a flow-based process enables the acquisition of intrinsic kinetic and mechanistic data essential to accelerate the design, optimization, and scale-up of RCF processes. Time-resolved product distributions and yields obtained from experiments with different catalyst loadings were used to identify and deconvolute events during solvolysis and reductive stabilization. Multi-bed RCF experiments provided unique insights into catalyst deactivation, and initial rate studies enabled the determination of kinetic parameters intrinsic to the solvolytic step of RCF. Overall, this work demonstrates the concept of flowthrough RCF, which will be vital for realistic scale-up of this promising approach.

530547 Hydrodeoxygenation of Sorbitol to Monofunctional Fuel Precursors over Co/TiO_2

Abstract:

Over the next two decades, demands for heavier fuels such as jet and diesel are projected to increase while demands for lighter fuels such as gasoline are projected to decrease.[1] Production of these heavy C_8-C_{22} hydrocarbons remains a challenge when using biomass feedstocks in large part because the monomeric units which comprise cellulose and hemicellulose possess only five or six carbon atoms. One potential route for producing these fuels involves converting carbohydrates into C_{5+} monofunctional oxygenates (MFs) by aqueous-phase hydrodeoxygenation (APHDO) followed by performing C-C coupling reactions.[2] APHDO consists of feeding a highly-oxygenated species in water at moderate temperatures (~500 K) along with hydrogen at high pressures (~6 MPa) over a bifunctional catalyst usually utilizing a noble metal on an acidic support.[3] In this chemistry it is desired to control the cleavage of C-O bonds and minimize the cleavage of C-C bonds by retro-aldol condensation, decarbonylation and decarboxylation reactions.[4] However, over-dehydration to lower-demand alkanes is undesired. In addition, the noble metals commonly used are costly, but utilization of low-cost base metals is often limited in APHDO due to their propensity to leach into the aqueous phase. However, recent developments have shown that Co/TiO_2 catalysts can be stabilized for aqueous-phase hydrogenations utilizing high-temperature pretreatments which take advantage of the strong metal-support interaction (SMSI).[5] The purpose of this study is to investigate the potential for MF production in sorbitol APHDO using a SMSI-stabilized Co/TiO_2 catalyst.[6]

Sorbitol APHDO of Co/TiO_2 produced a 56% yield (on a carbon basis) to MFs at a moderate WHSV of 0.70 h^{-1}. This is the highest yield reported from sorbitol in the literature. We estimate the Co catalyst is 50 to 1,000 times cheaper than previously reported precious metal catalysts. Increasing space time leads to excessive deoxygenation and therefore selectivity toward alkanes and CO_x while decreasing space time leads to higher oxygenates, including diols, oxygenated heterocycles, unidentified O_{3+} species. The MFs produced are mostly primary alcohols (37% at 0.70 h^{-1}), secondary alcohols (32%), and heterocycles (23%). Ketones (7%) and aldehydes (<1%) are produced in much lower yields. 67% of the MFs are C_{5+} and would therefore be suitable for coupling. FTICR-MS shows that sorbitol/anhydrosorbitol species can oligomerize at lower contact times to form C_9-C_{18} species which appear to break down to some degree at longer contact times to eventually form MFs. Overall, the product distribution vs. WHSV suggests that about one third of the fed carbon reacts through the retro-aldol route and overall 20% of the carbon is lost to C_1 gases such as CO_2 and methane. The need to suppress retro-aldol condensation and decarbonylation in sorbitol APHDO is a clear limitation for the production of C_{5+} MFs. In the present study, Co/TiO_2 catalyst shows irreversible deactivation with TOS due to sintering and leaching of the Co. Overall, this work shows that Co/TiO_2 can be used to produce monofunctional fuel precursors from sorbitol at high yields, but suffers from irreversible leaching and sintering due to chelation from highly-oxygenated species.

References

1. ExxonMobil, 2017 Outlook for energy: A view to 2040. ExxonMobil: Irving, 2017.
2. Kunkes, E. L.; Simonetti, D. A.; West, R. M.; Serrano-Ruiz, J. C.; Gärtner, C. A.; Dumesic, J. A., Catalytic conversion of biomass to monofunctional hydrocarbons and targeted liquid-fuel classes. *Science* **2008,** *322* (5900), 417-421.
3. Vilcocq, L.; Cabiac, A.; Especel, C.; Guillon, E.; Duprez, D., Transformation of sorbitol to biofuels by heterogeneous catalysis: Chemical and industrial considerations. *Oil & Gas Science and Technology–Revue d'IFP Energies nouvelles* **2013,** *68* (5), 841-860.
4. Li, N.; Huber, G. W., Aqueous-phase hydrodeoxygenation of sorbitol with Pt/SiO_2–Al_2O_3: Identification of reaction intermediates. *Journal of Catalysis* **2010,** *270* (1), 48-59.
5. Lee, J.; Burt, S. P.; Carrero, C. A.; Alba-Rubio, A. C.; Ro, I.; O'Neill, B. J.; Kim, H. J.; Jackson, D. H.; Kuech, T. F.; Hermans, I.; Dumesic, J. A.; Huber, G. W., Stabilizing cobalt catalysts for aqueous-phase reactions by strong metal-support interaction. *Journal of Catalysis* **2015,** *330,* 19-27.
6. Eagan, N. M.; Chada, J. P.; Wittrig, A. M.; Buchanan, J. S.; Dumesic, J. A.; Huber, G. W., Hydrodeoxygenation of Sorbitol to Monofunctional Fuel Precursors over Co/TiO2. *Joule* **2017,** *1* (1), 178-199.

533566 Catalytic Hydrogenation of Furfural over Rumop: Probing Bimetallic andÂ Compositional Effects on Catalyst Performance

Abstract:

Catalytic Hydrogenation of Furfural over RuMoP: Probing Bimetallic and Compositional Effects on Catalyst Performance

Yolanda Bonita, Timothy P. Oâ€™Connell, and Jason C. Hicks*

Department of Chemical and Biomolecular Engineering, University of Notre Dame, 182 Fitzpatrick Hall, Notre Dame, IN 46556

The synthesis of furfuryl alcohol from furfural uses an environmentally toxic copper chromite catalyst in the process. As possible alternatives, Cu-based catalysts such as supported Cu, bimetallic Cu-Ni, Cu-Co, Cu-Fe, and Cu-Ru have been studied to replace copper chromite. However, none of these catalysts surpass the high selectivity exhibited by copper chromite for furfural hydrogenation to furfuryl alcohol.

We have recently discovered a variety of novel transition metal phosphides that are highly selective to furfuryl alcohol and active at near ambient temperature. Contrary to the literature that suggested group VIII metals are not selective to furfuryl alcohol due to the binding configuration, we have synthesized group VIII metal (i.e. Ru) phosphides that exhibit high selectivity to furfuryl alcohol, similar to that of copper chromite and higher than other Cu-based catalysts. Moreover, when comparing pseudo-first order rate constants, we see an increase of > 5 times in the site-normalized, initial rate and 99% selectivity for bimetallic RuMoP. This phenomena was also observed in our previous studies (i.e. hydrodeoxygenation reaction)[1-4], which showcases bimetallic character and potential control of the catalytic properties based on catalyst composition. Indeed, we have observed direct influence on the activity and selectivity by varying the composition of $Ru_xMo_{2-x}P$, where the measured pseudo-first order rate constant was enhanced for Ru-rich materials (x = 1.2). In this presentation, we will present a thorough examination of bimetallic RuMoP as a furfural hydrogenation catalyst through detailed characterization and kinetic experiments.

References

1. Rensel, D. J., Rouvimov S., Gin M. E., Hicks J. C. Highly selective bimetallic FeMoP catalyst for C-O bond cleavage of aryl ethers. *J. Catal.* Sep 2013;305:256-263.
2. Rensel, D. J., Kim J., Bonita Y., Hicks J. C. Investigating the multifunctional nature of bimetallic FeMoP catalysts using dehydration and hydrogenolysis reactions. *Appl. Catal. A.* Aug 25 2016;524:85-93.
3. Rensel, D. J., Kim J., Jain V., Bonita Y., Rai N., Hicks J. C. Composition-directed $Fe_xMo_{2-x}P$ bimetallic catalysts for hydrodeoxygenation reactions. *Catal. Sci. Technol.* 2017;7:1857-1867.
4. Bonita, Y., Hicks J. C. Periodic trends from metal substitution in bimetallic Mo-based phosphides for hydrodeoxygenation and hydrogenation reactions. *J. Phys. Chem. C.* 2018, ASAP.

532119 Etherification of 5-Hydroxymethylfurfural Using Zeolite Catalysts

Abstract:

Biomass is a promising feedstock for the production of modern chemicals. In this regard, polymers represent an attractive class of potential biobased materials, and it may be possible to synthesize a variety of polymeric materials starting from 5-hydroxymethylfurfural (HMF). In this study, we show that H-BEA zeolite is highly active and selective for the production of a variety of HMF ethers which would be suitable for subsequent polymerization.

Several catalysts were evaluated for their etherification activity, including: H-BEA and H-ZSM-5 zeolites, Amberlyst-15 (A-15), and Î³-Al_2O_3. These catalysts were chosen to evaluate a variety of active sites and pore structures. Catalysts possessing BrÃ¸nsted acid sites (*i.e.*, H-ZSM-5 and H-BEA zeolites, and A-15) showed the highest selectivities to the ether product, with H-BEA achieving nearly quantitative selectivity. We hypothesize that the improved selectivity observed for H-BEA, compared with A-15, is due to confinement effects that inhibit both levulinate formation and biomolecular condensation (*i.e.*, resinification) reactions.

The rate of etherification is nearly constant for the series of ethanol, *n*-butanol, cyclohexanol, and phenol. Moreover, while the etherification rate saturates at high HMF concentrations, there is little variation in the rate across for alcohol concentrations ranging from 0.1 to 10 mol L^{-1}. These kinetics data, coupled with the invariance of the rate with varying alcohol data, suggest that the reaction proceeds through an oxocarbenium ion transition state that does not involve the etherifying alcohol.

532366 Mechanistic Insights into the Hydrogenolysis of Levoglucosanol over Bifunctional Platinum Silica-Alumina Catalysts in Tetrahydrofuran Solvent

Abstract:

The cellulose-derived platform molecule levoglucosenone (LGO) can be converted to a variety of high-value oxygenated chemicals. [1] LGO can be quantitatively hydrogenated to levoglucosanol (Lgol) over metal catalysts. [2] Herein, we report on the hydrogenolysis of Lgol to tetrahydrofurandimethanol (THFDM) and tetrahydropyran-2-methanol-5-hydroxyl (THP2M5H) over bifunctional platinum or palladium catalysts supported on silica-alumina in tetrahydrofuran solvent. [3,4] THFDM is an α,ω-diol and potential polymer precursor, while both THFDM and THP2M5H can be further upgraded to the commodity chemical 1,6-hexanediol.

A combination of experimental reaction kinetics studies and first principles calculations were used to investigate the mechanism and the roles of the metal and acid sites in this reaction. THFDM and THP2M5H are produced in parallel pathways, and the ketone intermediate THP2M5one is a precursor to THP2M5H. [3] The THFDM cis/trans ratio is independent of the Lgol threo/erythro ratio, suggesting that the mechanism passes through an acyclic intermediate which erases the stereochemistry of the Lgol reactant. ^{13}C radio-labeling was used to confirm the ring rearrangement forming THFDM. The catalyst undergoes moderate, reversible deactivation with time-on-stream in a continuous flow reactor. The reaction rates and product selectivities are comparable at 1.1% and 5.3% Pt loadings, indicating that the reaction rate is acid-limited at these metal loadings; the metal-catalyzed hydrogenation step is not rate-limiting (Figure 1). [4] The measured zero-order dependence in hydrogen indicates that the rate-limiting acid-catalyzed step precedes the fast hydrogenation step. As the Pt loading in Pt/SiAl catalysts is decreased, or when the bare SiAl support is separated from a Pt/SiO_2 catalyst in a dual-layer configuration, the selectivity towards identified products decreases (Figure 1). These results suggest that degradation reactions are avoided when the reactive intermediates formed over acid sites are rapidly hydrogenated over metal sites. First principles simulations were performed to investigate the energetics of the proposed reaction pathway consistent with the experimental results.

Figure 1. *Lgol consumption rate (left y-axis) and product selectivity (right y-axis) versus Pt loading for Pt/SiO_2-Al_2O_3 catalysts.*

These insights into the mechanism of Lgol hydrogenolysis provide a foundation for rationally designed processes to produce high-value chemicals from lignocellulosic biomass using bifunctional metal-acid catalysts.

References.

1. Cao, F.; Schwartz, T. J.; McClelland, D. J.; Krishna, S. H.; Dumesic, J. A.; Huber, G. W., Dehydration of cellulose to levoglucosenone using polar aprotic solvents. *Energy & Environmental Science* **2015**, *8* (6), 1808-1815.

2. He, J.; Huang, K.; Barnett, K. J.; Krishna, S. H.; Alonso, D. M.; Brentzel, Z. J.; Burt, S. P.; Walker, T.; Banholzer, W. F.; Maravelias, C. T.; Hermans, I.; Dumesic, J. A.; Huber, G. W., New catalytic strategies for [small alpha],[small omega]-diols production from lignocellulosic biomass. *Faraday Discussions* **2017**, *202* (0), 247-267.

3. Krishna, S. H.; McClelland, D. J.; Rashke, Q. A.; Dumesic, J. A.; Huber, G. W., Hydrogenation of levoglucosenone to renewable chemicals. *Green Chemistry* **2017**, *19* (5), 1278-1285.

4. Krishna, S. H.; Assary, R. S.; Rashke, Q. A.; Schmidt, Z. R.; Curtiss, L. A.; Dumesic, J. A.; Huber, G. W., Mechanistic Insights into the Hydrogenolysis of Levoglucosanol over Bifunctional Platinum Silica–Alumina Catalysts. *ACS Catalysis* **2018**, 3743-3753.

538548 Hydrothermal Stability of Chloromethyl Polystyrene Based Solid Acid Catalysts and Mechanism of Cellulose Hydrolysis

Abstract:

Cellulose is one of the major components of lignocellulosic biomass and the most abundant renewable polymer on earth. As such it could serve as a sustainable feedstock for the production of commercially important fuels and chemicals. However, cellulose conversion to smaller platform molecule is the major bottleneck for economically competitive utilization of lignocellulosic biomass.

The most common processes to depolymerize cellulose employ liquid acid or enzyme catalyzed hydrolysis to produce glucose. However, issues such as reactor corrosion, slow reaction rates, and inability to recycle limits the cost-effective conversion of cellulose. Alternatively, solid acids have been proposed as reusable catalysts for cellulose hydrolysis. Introduction of binding groups in the catalyst structure is hypothesized to increase the affinity between the solid acid and cellulose, brining the glycosidic bonds within contact with catalytic sites resulting in a higher hydrolysis rate. This hypothesis finds support in adsorption studies of small cellulosic molecules such as cellobiose and glucose. However, the mechanism of such catalytic activity has not been elucidated.

Chloromethyl polystyrene based bifunctional catalysts have showed high activity towards cellulose depolymerization. It is hypothesized the chlorine participates in hydrogen bonding with the hydroxyl groups in cellulose chains. However, the chloromethyl groups is rather label and could hydrolyze in hydrothermal conditions generating hydrochloric acid. Hydrochloric acid could also catalyze cellulose depolymerization, but its effect on the overall catalytic mechanism has not be evaluated.

In this work we studied partially sulfonated chloromethyl polystyrene catalyst for cellulose hydrolysis. Structural analysis of catalyst revealed that synthesis procedure results in little to no chlorine available on the external surface, but presence of hydroxyl groups, which can also participate in hydrogen bonding with cellulose. The catalyst showed high activity towards hydrolyzing cellulose resulting in levulinic acid and formic acid as major products. However, analysis of the solid and the liquid media revealed chloromethyl group hydrolysis and hydrochloric acid generation. This observation raised the question if the catalyst is indeed acting as a solid acid.

In order to elucidate the mechanism of hydrolysis we subjected the catalyst to hydrothermal reaction conditions in the absence of cellulose. Monitoring pH and chloride ion concentration, we developed a kinetic profile of hydrochloric acid generation. Cellulose hydrolysis kinetic models involve ion concentration as a preexponential factor on the expression for the reaction rate constant. We hypothesized that incorporating the time dependence of the acid concentration in the kinetic model of cellulose hydrolysis to glucose and subsequent dehydration to HMF, levulinic acid, and formic acid would be able to accurately predict experimental results. An acceptable agreement between model and experimental product yields pointed to the liquid acid acting as the primary catalyst, obscuring any catalytic activity, if any, of the solid catalyst. Further support was obtained by carrying out hydrolysis by a titrated solid acid catalyst, in which the sulfonic acid groups have been ion exchanged and, thus, rendered inactive; similar yields were obtained as the non-titrated catalyst. The results in this investigation highlight the need for systematic study of proposed solid acids for cellulose hydrolysis and consideration of potential release of homogeneous acids in the reaction media.

538054 Glycerol Transfer-Hydrogenation of Levulinic Acid Using Ru and Ir Carbene Organometallics Immobilized on Active Hydrotalcites

Abstract:

The mechanism and kinetics of levulinic acid (LA) hydrogenation to gamma-valerolactone (GVL) have been extensively studied over the last decade. Despite improvements in selectivity and the use of alternative hydrogen sources such as formic acid and secondary alcohols, economic production of GVL remains hindered by slow reaction rates and the high cost of current hydrogen sources. The use of glycerol, a waste byproduct in biodiesel synthesis, as a hydrogen donor could alleviate the economic challenges associated with GVL production. Herein we tested hydrogenation/deoxygenation of LA under dilute aqueous to neat conditions using Ru and Ir N-heterocyclic carbene (NHC) organometallic catalysts while also varying pH with potassium hydroxide. A total of 7 Ru and Ir homogeneous catalysts were screened for activity, the most active catalysts were then immobilized onto tunable hydrotalcites (HT). A 3:1 ratio of Mg:Al HT was used as a basis for all HT catalytic reactions while substitutions of Al with Fe, Ni, and Co were tested for improved catalytic activity. Products were identified using LCMS, HPLC, NMR, GCMS, and FTIR. The products formed and reaction rates were pH-dependent; acidic conditions favored ketal formation between LA and glycerol whereas alkaline conditions with an Ir(I) complex favored GVL and lactic acid formation at a turnover frequency (TOF) above 66,000 h^{-1} at 150ËšC for GVL. The Ir(I) catalyst achieved unprecedented reaction rates with glycerol, therefore, a comparison to reaction rates achieved with formic acid was demonstrated and showed negligible differences in rates and selectivity. Upon immobilizing the NHC catalysts on HT, the non-active Ru and Ir catalysts achieved activity over 10,000 h^{-1}. Under the same conditions the immobilized Ir(I)HT rate was reduced to 15,000 h^{-1} for GVL. The increase in activity from the other Ru and Ir(III) complexes tested are believed to be a result of the HT Lewis acid sites.

530511 Paired Electrocatalytic Hydrogenation and Oxidation of 5-Hydroxymethylfurfural for Efficient Production of Biorenewable Monomers

Abstract:

Electrochemical synthesis is a promising route for sustainable chemical production; however, its widespread application is hindered by economic limitations incurred from high capital and operating costs. In this presentation, we report highly efficient electrocatalytic conversion of 5-(hydroxymethyl)furfural (HMF) to biobased monomers by pairing HMF reduction and oxidation half reactions in a single electrochemical cell. HMF hydrogenation to 2,5-bis(hydroxymethyl)furan (BHMF) was achieved under mild electrolyte conditions and ambient temperature using a Ag/C cathode. Meanwhile, HMF oxidation to 2,5-furandicarboxylic acid (FDCA) with ~100% efficiency was facilitated under the same conditions by a homogeneous nitroxyl radical redox mediator, together with an inexpensive carbon felt anode. The selectivity and efficiency for Ag-catalyzed BHMF formation were sensitive to cathode potential, owing to competition from HMF hydrodimerization reactions and water reduction (hydrogen evolution). Moreover, the carbon support of Ag/C was active for HMF reduction and contributed to undesired dimer/oligomer formation at strongly cathodic potentials. As a result, high BHMF selectivity and efficiency were only achieved within a narrow potential range near â€"1.3 V. Fortunately, the selectivity of redox-mediated HMF oxidation was insensitive to anode potential, thus allowing HMF hydrogenation and oxidation half reactions to be performed together in a single cathode-potential-controlled cell. Electrocatalytic HMF conversion in a paired cell achieved high molar yields of BHMF and FDCA, and nearly doubled electron efficiency compared to the unpaired cells. Future research will be focused on rational design of an electrochemical flow reactor, in which the use of a membrane electrode assembly (MEA) can significantly lower the internal resistance and reduce the total energy demand.

525127 A Full FurfuralÂ Utilization over Ni/SiO_2 Catalysts

Abstract:

Furfural can be mass-produced from lignocellulose biomass and can be a platform chemical for producing valuable chemicals. In this study, we examine Ni/SiO_2 catalysts for the conversion of furfural under a hydrogen atmosphere. The reaction products include furfuryl alcohol (FA), furan, and propylene as major; butanal and other hydrogenated products, indicating the reaction routes involving side chain hydrogenation, decarbonylation, and ring opening. The reactivity and the product selectivity depend on the reaction temperature and the Ni particle size. The pretreatment by including calcination prior to hydrogen reduction leads to Ni/SiO_2â€"CR with large Ni particles (~ 15 nm) and a high selectivity to FA at below 200 °C. The Ni/SiO_2â€"R is hydrogen-pretreated without the calcination and it contains small Ni particles (~ 5 nm) and exhibits relatively high selectivity to furan. Moreover, the furan can be reacted seriously to produce propylene and CO by this Ni catalyst. The turnover frequency (TOF) of furfural conversion is 239 and 408 h^{-1}, respectively, on Ni/SiO_2-CR and Ni/SiO_2-R at 175 °C, when the former shows 100% selectivity to FA and the latter exhibits a selectivity of around 38% and 62%, to FA and furan, respectively. The results indicate that the furfural hydrotreating reaction over Ni catalysts is with structure sensitive and give a full furfural utilization, either to FA (a biofuel), or to propylene and CO (a chemical feedstock).

AUTHOR INDEX

AUTHOR INDEX

AUTHOR INDEX